Armin Lechleiter

Factorization Methods for Photonics and Rough Surfaces

Factorization Methods for Photonics and Rough Surfaces

by

Armin Lechleiter

universitätsverlag karlsruhe

Dissertation, Universität Karlsruhe (TH)
Fakultät für Mathematik, 2008

Impressum

Universitätsverlag Karlsruhe
c/o Universitätsbibliothek
Straße am Forum 2
D-76131 Karlsruhe
www.uvka.de

Universitätsverlag Karlsruhe 2008
Print on Demand

ISBN: 978-3-86644-256-6

Factorization Methods for Photonics and Rough Surfaces

Zur Erlangung des akademischen Grades eines

Doktors der Naturwissenschaften

von der Fakultät für Mathematik der
Universität Karlsruhe (TH)
genehmigte

Dissertation

von

Dipl. Math. Armin Lechleiter

aus Friedrichshafen

Tag der mündlichen Prüfung: 11. Juni 2008
Referent: Prof. Dr. Andreas Kirsch
Koreferent: PD Dr. Frank Hettlich

Abstract

We investigate several non-destructive testing problems for rough and periodic surfaces, where the task is to determine such a structure from scattered waves. We are interested in the analysis of the Factorization method applied to this special class of inverse scattering problems. To set the stage for the method, we show in the first chapter a range identity in factorizations which is suitable for application in the surface scattering context. In the same chapter it is further investigated how the method can be regularized in case of noisy data. These abstract results precede and are applied in three different case studies. First, we study a periodic transmission problem, motivated by identification problems in photonics. Second, we consider the problem of detecting a bounded contamination on a rough surface with Dirichlet boundary condition. Third, we continue the latter problem and consider the determination of a rough surface using near field scattering data.

Preface

The topic of this thesis is the Factorization method and its application to inverse problems in photonics and rough surface inverse scattering problems. Inverse scattering problems evolved into a self-contained research area 40 years ago and have extensively been studied since then. However, the focus of research has mainly been directed on inverse scattering problems involving bounded scattering objects. Recent progress in the analysis of direct scattering problems involving unbounded scatterers allows nowadays to have a closer look at the corresponding inverse problems.

One particular problem regarding the Factorization method for unbounded scatterers is that Sommerfeld's radiation condition has to be replaced by different radiation conditions, which cause standard assumptions of the method to fail. In the first chapter, we consequently provide an extension of the method's basic theorem on range identities in factorizations which is suitable for our purpose. This theorem on range identities has already some predecessors in the literature. The advantage of our version becomes evident when we use it to construct a Factorization method for the inverse medium problem, since eventual transmission eigenvalues of the direct scattering problem do no longer perturb the Factorization method. In Chapter I we also develop a regularization technique for the Factorization method in case of perturbed data, which is illustrated through numerical examples for an inverse medium scattering problem.

Chapter II is devoted to the study of an inverse problem arising in photonics: Monochromatic light is scattered by a periodic interface between two dielectrics and one aims to detect the interface from measurements of the scattered fields. This is a typical identification problem for photonic crystals and again we use the Factorization method to characterize the periodic interface explicitly in terms of the measurements. The technical difficulties of the analysis of the problem are heavy and partly caused by a lack of positivity and a large kernel of the middle operator in the Factorization.

In Chapter III we consider the detection of a bounded contamination on a rough surface from measurements of scattered fields a finite distance away from the contamination. It is an open problem to directly formulate a Factorization method for such measurements. We rely in this chapter on a factorization of an unphysical data operator and link that one afterwards to the physical measurement operator, improving results in [48].

Finally, Chapter IV presents the Factorization method as a tool to recover the entire rough surface and not just a bounded contamination. For various reasons, which will be discussed in detail, we have to restrict ourselves here to surfaces which have a surface elevation bounded by some constant divided by the wavenumber. Broadly speaking, this guarantees some necessary positivity assumption of the Factorization method to be satisfied. For completeness, we also explain in the end of this chapter why an approach in weighted spaces does not seem to extend this limited frequency range.

Note that parts of Sections I-5 and I-6 have been previously published in reference [57].

The work at this thesis has been partly supported by German Research Foundation DFG through a grant in the research training group GRK 1294 "Analysis, Simulation and Design of Nanotechnological Processes" at the Fakultät für Mathematik, Universität Karlsruhe (TH), and the financial support is gratefully acknowledged.

This thesis would not exist without support and encouragement of my colleagues at the Fakultät für Mathematik at Universität Karlsruhe (TH). First and foremost, I would like to thank my advisor Prof. Dr. Andreas Kirsch for many stimulating discussions and gentle supervision during the last years. Moreover, I am much obliged to PD Dr. Frank Hettlich for being the co-examiner of this thesis. Finally, it is a pleasure to thank Dr. Tilo Arens, Monika Behrens, Slavyana Geninska, Andreas Helfrich-Schkarbanenko, Sebastian Ritterbusch, Kai Sandfort, Susanne Schmitt, Arne Schneck and Dr. Henning Schon for much valuable help. Special thanks go to Dr. Tilo Arens and Dr. Henning Schon for their direct solvers for rough surface and medium scattering problems which are used to provide data for the inverse problems in Chapters I and III.

Contents

CHAPTER I

The Factorization Method and its Regularization

I-1. Introduction

In this first chapter we introduce the Factorization method in the context of a so-called inverse medium scattering problem and show how the method provides information on the location of an inhomogeneity in a background medium from measurements of scattered waves. We also show the theoretical basis of the method, see Theorem I.7, which is a certain range identity between operators linked by a factorization. It is this factorization which motivates the method's name. We also investigate the behavior of the method when only perturbed data are available and set up a simple regularization scheme for that situation. Finally, numerical examples for the inverse medium problem illustrate our theoretical results.

The Factorization method in inverse scattering theory has been introduced by Kirsch in [46], inspired by the Linear Sampling method developed by Colton and Kirsch in [21]. Both methods belong to the class of *qualitative methods* for inverse scattering problems. Their objective is often not to give a complete description of the scattering object but rather to determine certain features of the object, the most prominent one being certainly its shape (but, for example, not the precise values of the refractive index of an inhomogeneity). One might even imagine situations where one is merely interested in the number of connected components of the scattering object. A different approach is for instance usually taken by Newton-like methods [36, 34, 74]. Algorithms of this class linearize the non-linear inverse scattering problem locally and try to find all possible information about the scatterer. The potential of hybrid approaches combining Newton-like with qualitative methods does by the way not yet seem to be fully explored. Compared to Newton-like methods, two advantages of the Factorization method are its easy implementation and its independence of a-priori knowledge as for instance location or boundary conditions of the scatterer. Apart from the Linear Sampling method [12, 10], other qualitative methods are for instance the range test [73] or the notion of the scattering support [55, 32].

Applications of the Factorization method include inverse electromagnetic problems [47, 48] and inverse problems in elasticity [17] as well as inverse elliptic problems [47, 27], especially applications in impedance tomography [33, 35, 38] and optical tomography [39]. Let us also point out the applications of the method to scattering problems for periodic structures [7, 3], which will be considered in more detail in Chapter II. A striking feature of the method is its sound theoretical foundation together with its simple implementation. On the other hand, the assumptions necessary for the construction of a Factorization method are probably among the most restrictive ones compared to the methods cited above and there are many situations where it is not known yet whether or not the Factorization method applies, see [49]. Especially, the Linear Sampling method, closely related to the Factorization method through its methodology, has been shown to be applicable in much more situations than the Factorization method.

If it is applicable, the Factorization method provides an explicit binary criterion characterizing the shape of an obstacle or the support of an inhomogeneity. The data required for this shape characterization is the spectral decomposition of a certain operator which is computed from the measured data. For inverse scattering problems with far field measurements, this data is represented by the so-called far field operator. In this chapter, we are also interested in the behavior of the Factorization method when the measurements are only known approximately. This models the situation when only perturbed data with a certain error are at hand. Our results imply that the Factorization method still characterizes the scattering object in an asymptotic sense as the noise level goes to zero, if one regularizes the method in a suitable way. This result in turn validates application of the method to noisy data, but also offers new analytical ways to apply the Factorization method, compare [58]. We demonstrate this regularization technique for the special example of an inverse medium problem where one aims to detect the support of a bounded inhomogeneity of the refractive index from far field measurements.

I-2. Scattering by an Inhomogeneous Medium

In this chapter, we choose a two dimensional inverse medium scattering problem as a model problem to investigate regularization of the Factorization method with perturbed data. The corresponding direct problem is the scattering problem of incident acoustic waves by a bounded inhomogeneity in which the speed of sound differs from the speed of sound in the homogeneous and isotropic background medium. Our study of the two dimensional situation is no constraint of the Factorization method, indeed, everything in this chapter is also valid in three dimensions. However, we confine ourselves in all of the later chapters of this thesis to two dimensions and for consistency proceed in the same way here. Especially for the results in Chapters III and IV the extension to three dimensions is mathematically non-trivial, whereas that the results of Chapter II are independent of dimension.

Propagation of electromagnetic fields in three dimensions within linear, inhomogeneous and isotropic materials is described by the time dependent Maxwell's equations. If the time dependence of all fields involved is $\exp(-\mathrm{i}\omega t)$ for all times $t \in \mathbb{R}$ where $\omega > 0$ denotes the frequency, the time dependent Maxwell's equations reduce to the time harmonic ones. Assuming that there are no free currents these equations read

$$\mathrm{i}\omega \left(\varepsilon + \frac{\mathrm{i}\sigma}{\omega} \right) E + \nabla \times H = 0, \qquad -\mathrm{i}\omega\mu H + \nabla \times E = 0, \qquad \text{(I.1)}$$

where the positive electric permittivity ε and magnetic permeability μ as well as the non-negative conductivity σ are space dependent coefficients. We refer to [65] and [69] for a derivation of these equations from the time dependent Maxwell's equations. Note that, by taking the divergence of the last two equations, the well known identity $\nabla \cdot \nabla \times = 0$ yields $\nabla \cdot (\mu H) = 0$ and $\nabla \cdot ((\varepsilon + \mathrm{i}\sigma/\omega)E) = 0$. In case that all fields and coefficients are independent of the third variable, the curl of the electric and magnetic field is

$$\nabla \times E = \left(\frac{\partial E_3}{\partial x_2}, -\frac{\partial E_3}{\partial x_1}, \frac{\partial E_2}{\partial x_1} - \frac{\partial E_1}{\partial x_2}\right)^\top \quad \text{and}$$

$$\nabla \times H = \left(\frac{\partial H_3}{\partial x_2}, -\frac{\partial H_3}{\partial x_1}, \frac{\partial H_2}{\partial x_1} - \frac{\partial H_1}{\partial x_2}\right)^\top,$$

respectively. Plugging these two relations in the first and second equation of (I.1), we obtain two systems independent of the third variable of the magnetic and electric field,

$$\mathrm{i}\omega\left(\varepsilon + \frac{\mathrm{i}\sigma}{\omega}\right)E_1 + \frac{\partial H_3}{\partial x_2} = 0, \tag{I.2}$$

$$\mathrm{i}\omega\left(\varepsilon + \frac{\mathrm{i}\sigma}{\omega}\right)E_2 - \frac{\partial H_3}{\partial x_1} = 0, \tag{I.3}$$

$$\mathrm{i}\omega\left(\varepsilon + \frac{\mathrm{i}\sigma}{\omega}\right)E_3 + \frac{\partial H_2}{\partial x_1} - \frac{\partial H_1}{\partial x_2} = 0, \tag{I.4}$$

and

$$-\mathrm{i}\omega\mu H_1 + \frac{\partial E_3}{\partial x_2} = 0, \tag{I.5}$$

$$-\mathrm{i}\omega\mu H_2 - \frac{\partial E_3}{\partial x_1} = 0, \tag{I.6}$$

$$-\mathrm{i}\omega\mu H_3 + \frac{\partial E_2}{\partial x_1} - \frac{\partial E_1}{\partial x_2} = 0. \tag{I.7}$$

Note that the systems (I.2), (I.3) together with (I.7) and (I.4)-(I.6) decouple. Also, in the first and second of these systems of equations, the knowledge of E_3 and H_3 already determine the fields E and H. Now, plugging in (I.2) and (I.3) into (I.7) we find that H_3 satisfies the scalar equation

$$\nabla \cdot \left(\left(\frac{1}{\varepsilon + i\sigma/\omega}\right)\nabla u\right) + \omega^2 \mu u = 0. \tag{I.8}$$

Solutions to this equation are called *transverse electric modes*. Let us now additionally assume that the permeability μ equals some constant $\mu_0 > 0$

in all of $\mathbb{R}^2$. Plugging in (I.5) and (I.6) into (I.4) we find that E_3 satisfies

$$\Delta u + \omega^2 \mu_0 \left(\varepsilon + \frac{\mathrm{i}\sigma}{\omega} \right) u = 0. \tag{I.9}$$

Solutions of the latter equations are called *transverse magnetic modes.*

In the remainder of this work we will only study equation (I.9), which is known as the *Helmholtz equation.* We will also assume in this entire chapter that ε equals some constant $\varepsilon_0 > 0$ outside of some bounded domain and that σ has compact support. More precisely, we assume that the support D of $\varepsilon_0 - \varepsilon$ is bounded, that the complement $\mathbb{R}^2 \setminus \overline{D}$ is connected and that $\operatorname{supp}(\sigma) \subset \overline{D}$. The *wavenumber* k is defined by $k := \omega\sqrt{\varepsilon_0 \mu_0} > 0$ and the *refractive index* n by

$$n := \frac{1}{\sqrt{\varepsilon_0}} \left(\varepsilon + \mathrm{i}\frac{\sigma}{\omega} \right)^{1/2}.$$

In case of a complex refractive index (with non-negative imaginary part), the square root in the latter definition is the analytic extension of the square root function to the slit complex plane with branch cut along the negative imaginary axis. Since $n = 1$ outside of D it will also be convenient to deal with the *contrast*

$$q := n^2 - 1,$$

which has compact support. We can then restate the Helmholtz equation as

$$\Delta u + k^2(1+q)u = 0 \qquad \text{in } \mathbb{R}^2. \tag{I.10}$$

In the remainder of this section, we study scattering of an *incident plane wave*

$$u^i(x,\theta) = e^{\mathrm{i}k\, d\cdot x}, \qquad x \in \mathbb{R}^2, \tag{I.11}$$

of direction $d \in \mathbb{S}^1 := \{\theta \in \mathbb{R}^2 : \ |\theta| = 1\}$ at the inhomogeneity supported in D. Obviously, $u^i(\cdot, d)$ satisfies the equation $\Delta u + k^2 u = 0$ with constant coefficients in $\mathbb{R}^2$. Hence, the incident field u^i causes a scattered field u^s such that the total field

$$u^t := u^i + u^s$$

satisfies the Helmholtz equation (I.10) in all of $\mathbb{R}^2$. Moreover, generation of the scattered field u^s is a local effect in D and therefore we impose that the scattered field u^s propagates away from D. This property is expressed as a *radiation condition*, namely the Sommerfeld radiation condition, which states that

$$\lim_{r\to\infty} \sqrt{r}\left(\frac{\partial u^s}{\partial r} - iku^s\right) = 0, \qquad r = |x|, \tag{I.12}$$

where the limit is uniform in all directions $x/|x| \in \mathbb{S}^1$. Roughly speaking, the Sommerfeld radiation condition imposes on u^s to behave as a spherical wave propagating away from the scatterer D. More precisely, compare [11, Section 4.1], any solution u of the Helmholtz equation which satisfies (I.12) has the asymptotic behavior

$$u(x) = \frac{e^{ik|x|}}{\sqrt{|x|}}\left(u_\infty(\hat{x}) + \mathcal{O}\left(\frac{1}{|x|}\right)\right), \qquad |x| \to \infty, \tag{I.13}$$

where the limit is again uniform in all directions $\hat{x} := x/|x| \in \mathbb{S}^1$. The function $u_\infty \in L^2(\mathbb{S}^1)$, which describes the leading order behavior of the radial part of $u(x)$ for $|x| \to \infty$ is known as the *far field pattern* of u. Probably the most important radiating solution of the Helmholtz equation is the radiating fundamental solution Φ of the Helmholtz equation. In two dimensions,

$$\Phi(x,y) = \frac{\mathrm{i}}{4} H_0^{(1)}(k|x-y|), \qquad x \neq y, \tag{I.14}$$

where $H_0^{(1)}$ denotes the Hankel function of the first kind of order zero, see, e.g., [70]. Recall that the far field pattern of $\Phi(\cdot, y)$, which is also called a *point source* at y, is denoted by $\Phi_\infty(\cdot, y)$ and given by (see, e.g., [11, Section 4.1])

$$\Phi_\infty(\hat{x}, y) = \frac{e^{\mathrm{i}\pi/4}}{\sqrt{8\pi k}} e^{-\mathrm{i}k\,\hat{x}\cdot y}, \qquad \hat{x} \in \mathbb{S}^1,\ y \in \mathbb{R}^2. \tag{I.15}$$

Since the incident and the total field satisfy two different Helmholtz equations, we find that the scattered field solves

$$\Delta u^s + k^2(1+q)\,u^s = -k^2 q\, u^i \qquad \text{in } \mathbb{R}^2.$$

We will solve this equation for a wider class of right hand sides and consider the equation

$$\Delta v + k^2(1+q)\,v = -k^2 q\, f \qquad \text{in } \mathbb{R}^2, \tag{I.16}$$

for $f \in L^2(D)$, where we seek again for a solution which also satisfies the Sommerfeld radiation condition. Moreover, we require u and its normal derivative to be continuous over interfaces where the contrast q jumps. Since we consider $f \in L^2(D)$, we understand the last equation in the weak sense: formal integration of (I.16) against a testfunction $\phi \in \mathcal{D}(\mathbb{R}^2)$ and an application of Green's first identity shows that

$$\int_{\mathbb{R}^2} \left(\nabla v \nabla \overline{\phi} - k^2(1+q)\, v\overline{\phi}\right) \mathrm{d}x = k^2 \int_D q f \,\overline{\phi}\, \mathrm{d}x \qquad \text{for all } \phi \in \mathcal{D}(\mathbb{R}^2). \tag{I.17}$$

Hence we seek for a (weak) solution $v \in H^1_{\mathrm{loc}}(\mathbb{R}^2)$ which satisfies the weak formulation (I.17) and Sommerfeld's radiation condition (I.12). It is easy to observe by Cauchy's inequality that for $v \in H^1_{\mathrm{loc}}(\mathbb{R}^2)$, all integrals in (I.17) are well defined. Moreover, well known regularity results for elliptic equations [63, Theorem 4.16] and Sobolev's embedding theorem [28, Section 7.7] show that the field v which solves (I.17) is continuously differentiable outside of D such that the radiation condition (I.12) is well defined for v.

We treat equation (I.16) using the Lippmann-Schwinger integral equation approach. Therefore we recall that the volume potential

$$\mathcal{V}f = \int_D \Phi(\cdot, y) f(y)\, \mathrm{d}y, \qquad x \in \mathbb{R}^2, \tag{I.18}$$

is a bounded operator from $L^2(D)$ into $H^2_{\mathrm{loc}}(\mathbb{R}^2)$, see for instance [63, Equation (6.10)] and the references therein.

THEOREM I.1. *Let* $q \in L^\infty(D)$ *such that* $1 + \mathrm{Re}\,(q) > 0$ *and* $\mathrm{Im}\,(q) \geq 0$ *and* $f \in L^2(D)$.

(a) If a function $v \in H^1_{\mathrm{loc}}(\mathbb{R}^2)$ *solves* (I.16) *and* (I.12), *then the restriction* $v|_D$ *solves the following* Lippmann-Schwinger equation

$$v = k^2\, \mathcal{V}(q(v+f))|_D \,. \tag{I.19}$$

If $v \in L^2(D)$ *solves this Lippmann-Schwinger equation, then* $k^2\mathcal{V}(q(v+f))$ *provides an extension of* v *to a function in* $H^1_{\mathrm{loc}}(\mathbb{R}^2)$ *which solves* (I.16) *and* (I.12).

(b) The Lippmann-Schwinger equation (I.19) *has a unique solution for all* $k > 0$ *and* $f \in L^2(D)$, *which depends continuously on* f.

Proof. (a) Assume that v solves (I.16) together with (I.12). It is well known [63, Section 9] that Φ is a fundamental solution for the Helmholtz operator $-\Delta - k^2$, that is, $\Delta\Phi(\cdot, y) + k^2\Phi(\cdot, y) = -\delta_x$ where δ_x denotes the Dirac distribution in x, $\delta_x(\phi) = \phi(x)$ for $\phi \in \mathcal{D}(\mathbb{R}^2)$. Therefore, see [63, Page 197], $\Delta\mathcal{V}f + k^2\mathcal{V}f = -f$ holds in the distributional sense for all $f \in \mathcal{E}'(\mathbb{R}^2)$, thus, especially for all functions $f \in L^2(D)$. For such f, the function $\mathcal{V}f$ belongs to $H^2_{\text{loc}}(\mathbb{R}^2)$ and the equality $\Delta\mathcal{V}f + k^2\mathcal{V}f = -f$ holds in $L^2(\mathbb{R}^2)$. In consequence,

$$\Delta\mathcal{V}(k^2q(v+f)) + k^2\mathcal{V}(k^2q(v+f)) = -k^2q(v+f)$$

and a simple computation shows that $w = v - \mathcal{V}(k^2q(v+f))$ is a solution of the Helmholtz equation $\Delta w + k^2 w = 0$ with zero right hand side. Moreover, w satisfies the Sommerfeld radiation condition and is hence an entire radiating solution of the Helmholtz equation. It is well known that such a function vanishes identically in all of $\mathbb{R}^2$, compare [22].

Assume now that $v \in L^2(D)$ solves the Lippmann-Schwinger equation (I.19). We extend v by the volume potential on the right-hand side of (I.19) to a radiating solution of the Helmholtz equation in $\mathbb{R}^2$. As mentioned above, $\mathcal{V}$ is a bounded operator from $L^2(D)$ into $H^2_{\text{loc}}(\mathbb{R}^2)$, thus, $v \in H^2_{\text{loc}}(\mathbb{R}^2)$. We computed in the first part of the proof that $\Delta\mathcal{V}(k^2q(v+f)) + k^2\mathcal{V}(k^2q(v+f)) = -k^2q(v+f)$ in $\mathbb{R}^2$. From the latter equation combined with the Lippmann-Schwinger equation it follows that $\Delta v + k^2 v = -k^2q(v+f)$, which means that $\Delta v + k^2(1+q)v = -k^2qf$. The proof of this part is complete.

(b) The boundedness of $\mathcal{V}$ from $L^2(D)$ into $H^2_{\text{loc}}(\mathbb{R}^2)$ implies that $f \mapsto k^2\, \mathcal{V}(qf)|_D$ is a compact operation on $L^2(D)$. Writing the Lippmann-Schwinger equation (I.19) in the form

$$v - k^2\, \mathcal{V}(qv)|_D = k^2\, \mathcal{V}(qf)|_D\,,$$

one observes that Riesz-Fredholm theory, see, e.g., [63, Theorem 2.19], implies that the Lippmann-Schwinger equation is solvable for any right-hand side if its only solution for $f = 0$ is the trivial solution. Hence, to finish the proof we need to show injectivity of $\mathrm{Id} - k^2\, \mathcal{V}(q\,\cdot)|_D$ on $L^2(D)$.

Assume that $v - k^2\, \mathcal{V}(qv)|_D = 0$. Then, according to part (a) of the proof, the extension of v by $k^2\mathcal{V}(qv)$ to all of $\mathbb{R}^2$ is a radiating solution of (I.16) for zero right hand side. The mapping properties of $\mathcal{V}$ imply that $v \in H^2_{\text{loc}}(\mathbb{R}^2)$. Using this regularity result, we conclude by Green's first identity [63] in the domain $\Omega_R := \{x \in \mathbb{R}^2 : |x| > R\}$, $R > 0$ large enough, that

$$\int_{\Omega_R} \left(|\nabla v|^2 - k^2(1+q)|v|^2 \right) \mathrm{d}x = \int_{\partial\Omega_R} \overline{v} \frac{\partial v}{\partial \nu} \, \mathrm{d}s.$$

The imaginary part of the left hand side is non-positive, since by assumption $\mathrm{Im}\,(q) \geq 0$. Hence,

$$\mathrm{Im} \left(\int_{\partial\Omega_R} v \frac{\partial \overline{v}}{\partial \nu} \, \mathrm{d}s \right) \geq 0,$$

and Rellich's Lemma implies (see, e.g., [11, Theorem 3.6] for the argument) that v vanishes in Ω_R for R large enough. However, in this situation the unique continuation property, which holds in two dimensions for elliptic equations with bounded measurable coefficients, see [79], implies that v vanishes in all of $\mathbb{R}^2$. □

Combining parts (a) and (b) of the latter theorem, we obtain the following corollary.

COROLLARY I.2. *For all* $k > 0$ *and* $q \in L^\infty(D)$ *such that* $1 + \mathrm{Re}\,(q) > 0$ *and* $\mathrm{Im}\,(q) \geq 0$, *there is a unique solution* $v \in H^2_{\mathrm{loc}}(\mathbb{R}^2)$ *of the scattering problem* (I.16), (I.12). *Especially, for all incident plane waves* $u^i(\cdot, \theta)$ *of the form* (I.11) *there is a unique scattered field* $u^s(\cdot, \theta) \in H^2_{\mathrm{loc}}(\mathbb{R}^2)$ *which satisfies* (I.16) *with* $f = u^i(\cdot, \theta)$ *and the Sommerfeld radiation condition* (I.12).

I-3. A Factorization Method for the Inverse Medium Problem

After the presentation of the solution theory for the direct scattering problem in an inhomogeneous medium, we can now state the inverse problem we want to investigate: Given the far field pattern $u^s_\infty(\hat{x}, \theta)$ of the scattered field $u^s(\cdot, \theta)$ for all angles $\hat{x} \in \mathbb{S}^1$ and all incident directions $\theta \in \mathbb{S}^1$, determine the support D of the contrast q! We are going to develop a Factorization method for this problem which follows the lines of construction in [49, Chapter 4]. However, using a refined version of the method's basic theorem on range identities, our version has the advantage that the transmission eigenvalues of the direct scattering problem (see, e.g., [49, Chapter 4] for a definition) do not enter the analysis of the inverse problem at all.

Regarding the contrast q we have to strengthen our assumptions from the last section to construct the Factorization method. Even if the factorization of the far field operator in Theorem I.4 itself does hold for bounded and measurable contrasts, we are not able to show the crucial range identity of the method, see Theorem I.7, in such a general situation. We therefore require in the remainder of this chapter the following assumptions for q.

ASSUMPTION I.3. *The contrast* $q \in L^\infty(D)$ *satisfies* $\operatorname{Im} q \geq 0$ *and* $q \neq 0$ *almost everywhere in* D*. Moreover, we impose that*

$$\operatorname{Re}(q) \geq c|q| \qquad \text{almost everywhere in } D \tag{I.20}$$

for some constant $c > 0$.* *Finally, one of the following two assumptions needs to be satisfied: Either*

$$\text{for every closed ball } B \subset D \text{ there is } C_B > 0 \text{ such that } |q| > C_B \text{ in } B, \tag{I.21}$$

or

$$\text{there is } \varepsilon > 0 \text{ such that } \int_{D_\varepsilon} \frac{1}{|q|} \, \mathrm{d}x < \infty, \quad \text{where } D_\varepsilon = \{x \in D : \operatorname{dist}(x, \partial D) < \varepsilon\}. \tag{I.22}$$

Note that Assumption I.3 allows the contrast q to vanish on the boundary ∂D. For instance, any real contrast $q \in L^\infty(D)$ which is strictly positive on any compact subset of D satisfies (I.20) with $c = 1$ and also (I.21). If the absolute value of the contrast is not bounded from below on any compact subset of D, then q must to decay slowly near the boundary such that $1/|q|$ is integrable in some neighborhood of ∂D and (I.22) holds. An example for such a contrast is, for instance, the function $q(x) = |x|$ in the domain $D = \{x \in \mathbb{R}^2 : |x| < 1\}$. This contrast does not satisfy (I.21), but it does satisfy (I.22).

*The letters C and c denote generic constants in the entire work. Their numerical values might change from one occurrence to the other. To emphasize the dependence of the constant on some parameter p we write $C(p)$.

Let us next introduce the *far field operator*

$$F : L^2(\mathbb{S}^1) \to L^2(\mathbb{S}^1), \qquad g \mapsto \int_{\mathbb{S}^1} g(\theta)\, u_\infty(\cdot,\theta)\,\mathrm{d}s(\theta).$$

The far field operator is an integral operator with continuous (and even analytic) kernel $u_\infty(\cdot,\cdot)$, see, e.g, [11, Section 4.1], and is therefore compact on $L^2(\mathbb{S}^1)$. By linearity of the scattering problem, Fg which maps a density g to the far field corresponding to the incident field v_g, defined by

$$v_g(x) = \int_{\mathbb{S}^1} g(\theta) u^i(\cdot,\theta)\,\mathrm{d}s(\theta) = \int_{\mathbb{S}^1} g(\theta) e^{\mathrm{i}k\,\theta\cdot x}\,\mathrm{d}s(\theta), \qquad x \in \mathbb{R}^2.$$

The function v_g is called the Herglotz wave function of g. If we are given far field measurements $\{u_\infty(\hat{x},\theta) : \hat{x},\theta \in \mathbb{S}^1\}$ for all angles $\hat{x}$ and incident directions θ, then we also know the far field operator (and vice versa). Therefore we investigate in the remainder of this section the following inverse problem:

Given F, determine the support D of the contrast q!

The basis of the Factorization method is a suitable factorization of F. To state this factorization, we still need to define the Herglotz operator,

$$H : L^2(\mathbb{S}^1) \to L^2(D), \quad g \mapsto \sqrt{|q(x)|} \int_{\mathbb{S}^1} g(\theta) e^{\mathrm{i}k\,\theta\cdot x}\,\mathrm{d}s(\theta), \quad x \in D.$$

Note that Hf is the restriction of the Herglotz wave function v_g to D (multiplied by $\sqrt{|q|}$). The Cauchy-Schwartz inequality implies that H is a bounded linear operator. The adjoint of H with respect to the inner product of $L^2(D)$ is easily computed to be

$$H^* : L^2(D) \to L^2(\mathbb{S}^1), \quad f \mapsto \int_D f(x)\sqrt{|q(x)|} e^{-\mathrm{i}k\,\theta\cdot x}\,\mathrm{d}x, \quad \theta \in \mathbb{S}^1.$$

THEOREM I.4. *The far field operator can be factored as*

$$F = \gamma H^* T H,$$

where $\gamma = \exp(\mathrm{i}\pi/4)/\sqrt{8\pi k}$ *and* T *is defined as*

$$Tf = k^2 \frac{q}{\sqrt{|q|}} \left(\frac{1}{\sqrt{|q|}} f + v|_D \right),$$

and $v \in H^1_{\mathrm{loc}}(\mathbb{R}^2)$ *is the radiating solution of*

$$\Delta v + k^2(1+q)\, v = -k^2 \frac{q}{\sqrt{|q|}} f \qquad \text{in } \mathbb{R}^2. \tag{I.23}$$

Proof. Recall from Theorem I.1 in Section I-2 that the Helmholtz equation (I.23) together with the Sommerfeld radiation condition (I.12) has always a unique solution $v \in H^1_{\mathrm{loc}}(\mathbb{R}^2)$, which can be represented by a volume potential, see (I.19),

$$v = k^2 \mathcal{V}\left(\frac{q}{\sqrt{q}} f + q\, v|_D\right). \tag{I.24}$$

We define the *data-to-pattern operator* $G : L^2(D) \to L^2(\mathbb{S}^1)$ by $f \mapsto v_\infty$, where v_∞ is the far field pattern of v solving (I.23). Then the equality $F = GH$ holds by construction. The adjoint H^* maps $f \in L^2(D)$ to the far field pattern of the volume potential $\mathcal{V}(\sqrt{|q|} f)/\gamma$, $\gamma = \exp(\mathrm{i}\pi/4)/\sqrt{8\pi k}$, since the far field pattern of the fundamental solution $x \mapsto \Phi(x,y)$ is $\gamma \exp(-\mathrm{i}k\, \hat{x} \cdot y)$, compare (I.15). Thus, $\gamma H^* f = (\mathcal{V}(\sqrt{|q|} f))_\infty$. Replacing now f by Tf as defined above, we find

$$\gamma H^* T f = k^2 \left(\mathcal{V}\left(\frac{q}{\sqrt{|q|}} f + q\, v|_D\right)\right)_\infty .$$

Comparing the right-hand side of the latter equation with (I.24) shows that $\gamma H^* T f$ is the far field pattern of the radiating solution of (I.23), that is, $\gamma H^* T = G$. Therefore we conclude that

$$F = \frac{e^{\mathrm{i}\pi/4}}{\sqrt{8\pi k}} H^* T H.$$

□

The Herglotz wave operator H, the middle operator T and also the data-to-pattern operator G, introduced in the proof of the last theorem, have certain special properties which we will exploit for construction of the Factorization method. For convenience, we announce these properties in the following lemma.

LEMMA I.5. *Assume that the contrast q satisfies Assumption I.3. Then the following statements hold.*

(a) The Herglotz operator $H : L^2(\mathbb{S}^2) \to L^2(D)$ is injective.

(b) The middle operator $T : L^2(D) \to L^2(D)$ is injective and has a natural splitting in a sum of a coercive and a compact operator:

$$T f = k^2 \frac{q}{|q|} f + k^2 \frac{q}{\sqrt{|q|}} v|_D = T_0 f + T_1 f$$
$$\text{with } T_0 f := k^2 \frac{q}{|q|} f \text{ and } T_1 f := k^2 \frac{q}{\sqrt{|q|}} v|_D ,$$

where v is again the radiating solution of (I.23). *Moreover, T is an isomorphism and*

$$\operatorname{Im}\left(\int_D \overline{f}\, T f \,\mathrm{d}x \right) \geq 0 \qquad f \in L^2(D). \tag{I.25}$$

(c) The data-to-pattern operator $G : L^2(D) \to L^2(\mathbb{S}^1)$ characterizes D in the following way: a point $z \in \mathbb{R}^2$ belongs to D if and only if the far field pattern $\Phi_\infty(\cdot, z)$ of a point source at z belongs to the range of G. The same holds for the adjoint H^* of the Herglotz operator.

Recall that an operator $T : L^2(D) \to L^2(D)$ is, by definition, coercive, if there is $c > 0$ such that $\operatorname{Re} \int_D \overline{f}\, T f \,\mathrm{d}x > c\|f\|^2_{L^2(D)}$ for all $f \in L^2(D)$. For a selfadjoint operator, this corresponds to the notion of positive definiteness: we call a selfadjoint operator $T : L^2(D) \to L^2(D)$ positive definite, if there is $c > 0$ such that $\int_D \overline{f}\, T f \,\mathrm{d}x > c\|f\|^2_{L^2(D)}$ for all $f \in L^2(D)$. Moreover, a selfadjoint operator $T : L^2(D) \to L^2(D)$ is called positive, if $\int_D \overline{f}\, T f \,\mathrm{d}x > 0$ for all $0 \neq f \in L^2(D)$ and non-negative or positive semidefinite, if $\int_D \overline{f}\, T f \,\mathrm{d}x \geq 0$ for all $f \in L^2(D)$. All these definitions have of course natural extensions to general Hilbert spaces.

Proof. (a) If Hg vanishes on D, then

$$x \mapsto \int_{\mathbb{S}^1} g(\theta) e^{\mathrm{i}k\,\theta\cdot x} \,\mathrm{d}s(\theta)$$

is an entire solution of the Helmholtz equation which vanishes on D. The unique continuation property directly implies that Hg vanishes entirely in $\mathbb{R}^2$ and the Jacobi-Anger expansion [11, Section 3.3] shows that g vanishes.

(b) If is easily seen that the part T_1 of T is compact: since the solution v of the scattering problem (I.23) restricted to D belongs to $H^2(D)$, compactness of the embedding $H^2(D) \hookrightarrow L^2(D)$ implies compactness of $L^2(D) \ni f \mapsto v|_D \in L^2(D)$. Moreover, multiplication by $k^2 \frac{q}{\sqrt{|q|}}$ is a bounded operation on $L^2(D)$ and hence T_1 is compact. For the part T_0 we note that Assumption I.3 implies

$$\mathrm{Re}\left(\int_D \overline{f}\, T_0 f \,\mathrm{d}x\right) = k^2 \int_D \frac{\mathrm{Re}\,(q)}{|q|} |f|^2 \,\mathrm{d}x \geq ck^2 \|f\|^2_{L^2(D)},$$

that is, T_0 is coercive on $L^2(D)$.

Next we prove that T is injective. Assume that $Tf = 0$ which means that $f + \sqrt{|q|}v = 0$ in D and rewrite the last equation as $-f/\sqrt{|q|} = v$. Since v is a radiating solution of $\Delta v + k^2(1+q)\,v = -k^2 qf/\sqrt{|q|}$, we find that $\Delta v + k^2(1+q)\,v = k^2 qv$ in $\mathbb{R}^2$. Clearly, we obtain that v is an entire radiating solution of $\Delta v + k^2 v = 0$ in all of $\mathbb{R}^2$. Since the latter problem is uniquely solvable for any right hand side in $L^2(D)$, we conclude that v vanishes and hence f vanishes, too, implying that T is injective. Thus, T in an injective Fredholm operator of index zero since it is the sum of a coercive and a compact operator. Consequently, T is an isomorphism.

Finally, we show (I.25). For $f \in L^2(D)$,

$$\int_D \overline{f}\, Tf \,\mathrm{d}x = k^2 \int_D \frac{q}{|q|} w\overline{f} \,\mathrm{d}x = k^2 \int_D \frac{q}{|q|}\left(|w|^2 - \frac{q}{\sqrt{|q|}} w\overline{v}\right) \mathrm{d}x, \quad \text{(I.26)}$$

with $w := f + \sqrt{|q|}\; v|_D$. From the definition of v in (I.23) we observe that

$$\Delta v + k^2 v = -k^2 \frac{q}{\sqrt{|q|}} f - k^2 qv = -k^2 \frac{q}{\sqrt{|q|}} w.$$

We plug this relation into (I.26) and find by Green's first identity in D

$$\begin{aligned}\int_D \overline{f}\, Tf \,\mathrm{d}x &= k^2 \int_D \frac{q}{|q|} |w|^2 \,\mathrm{d}x + \int_D (\Delta v + k^2 v)\,\overline{v} \,\mathrm{d}x \\ &= k^2 \int_D \frac{q}{|q|} |w|^2 \,\mathrm{d}x + \int_D \left(k^2 |v|^2 - |\nabla v|^2\right) \mathrm{d}x + \int_{\partial D} \frac{\partial v}{\partial \nu} \overline{v} \,\mathrm{d}s.\end{aligned}$$

Note that in the last integral on the right, ν denotes the exterior unit normal to D. Since the imaginary part of q is by assumption non-negative, it is sufficient to show that $\mathrm{Im}\left(\int_{\partial D} \overline{v}\, \partial v/\partial\nu \,\mathrm{d}s\right) \geq 0$ to conclude that $\mathrm{Im}\left(\int_D \overline{f} T f \,\mathrm{d}x\right) \geq 0$. However, assuming that $\mathrm{Im}\left(\int_{\partial D} \overline{v}\, \partial v/\partial\nu \,\mathrm{d}s\right) < 0$ Rellich's lemma [11, Theorem 3.6] yields that v vanishes entirely, hence $\mathrm{Im}\left(\int_{\partial D} \overline{v}\, \partial v/\partial\nu \,\mathrm{d}s\right) = 0$, contradiction.

(c) It is not difficult to see that the far field of a point source at $z \in \mathbb{R}^2 \backslash \overline{D}$ cannot belong to the range of G: Assume that u is a radiating solution of (I.23) for some $f \in L^2(D)$ with far field $\Phi_\infty(\cdot, z)$ in $\mathbb{R}^2$. Hence, the far field pattern of the radiating solution $w = u - \Phi(\cdot, z)$ vanishes and the asymptotic behavior (I.13) implies that

$$\begin{aligned}\int_{|x|=R} |w|^2 \,\mathrm{d}s &= \frac{1}{R} \int_{|x|=R} |w^\infty(\hat{x})|^2 \,\mathrm{d}s + \frac{1}{R} \int_{|x|=R} \mathcal{O}\left(\frac{1}{R^2}\right) \mathrm{d}s \\ &= 2\pi \int_{\mathbb{S}^1} |w^\infty|^2 \,\mathrm{d}s + \mathcal{O}\left(\frac{1}{R^2}\right) = \mathcal{O}\left(\frac{1}{R^2}\right)\end{aligned}$$

and hence $\int_{|x|=R} |w|^2 \,\mathrm{d}s \to 0$ as $R \to \infty$. Rellich's Lemma [11, Theorem 3.5] and an analytic continuation argument imply that w vanishes in the exterior of $D \cup \{y\}$, that is, $\Phi(\cdot, z) = u$ in $\mathbb{R}^2 \setminus (D \cup \{y\})$. However, $\Phi(\cdot, z)$ has a singularity at $z \notin D$ but u belongs by Theorem I.1 to $H^2_{\mathrm{loc}}(\mathbb{R}^2)$. Hence, by the Sobolev embedding theorem [28, Corollary 7.11], u is a continuous function in $\mathbb{R}^2$ and cannot have a singularity at y, contradiction.

For $z \in D$ we need to show the existence of some ϕ such that $H^*\phi = \Phi^\infty(\cdot, y)$. Under assumption (I.21), such an element is constructed literally as in [49, Theorem 4.6]. Therefore we only consider assumption (I.22) here: There is $\varepsilon > 0$ such that $\int_{D_\varepsilon} 1/|q| \,\mathrm{d}x < \infty$ where $D_\varepsilon = \{x \in D : \mathrm{dist}(x, \partial D) < \varepsilon\}$. For $z \in D$ there is a smooth function $\chi \in C_c^\infty(\mathbb{R}^2)$ such that $\chi = 0$ in a neighborhood of z and $\chi = 1$ in a neighborhood of $\mathbb{R}^2 \setminus D$. We can moreover choose χ such that $\mathrm{supp}(\chi) \cap D$ is included in D_ε. The far field of $\Phi(\cdot, z)$ equals the far field of $w = \chi\Phi(\cdot, z)$. Moreover, the function

$$f = -\frac{1}{k^2}\left(\Delta w + k^2(1+q)\, w\right) \frac{\sqrt{|q|}}{q}$$

is square integrable in D due to assumption (I.22). For this f, the radiating solution of (I.23) equals w outside of D, by the unique solvability of the latter equation. Hence, the far field of w belongs by definition to the range of G, and by construction the far field of $\Phi(\cdot, y)$ belongs to this range, too. According to part (b), T is an isomorphism. Since $\gamma H^*T = G$, the range of G equals the range of H^* and hence the proof is complete. □

The preparations made in the last lemma combined with an abstract result on range equalities, see Theorem I.7, directly yields a characterization of D in terms of F. For simplicity, we first state this characterization in the next theorem and postphone the required result on range identities in the following Section I-4. As a prerequisite, we introduce real and imaginary part of a bounded linear operator T on a Hilbert space, which are defined in accordance with the corresponding definition for complex numbers,

$$\operatorname{Re} T := \frac{1}{2}(T + T^*), \qquad \operatorname{Im} T := \frac{1}{2\mathrm{i}}(T - T^*).$$

THEOREM I.6. *Let* $(\lambda_j, \psi_j)_{j\in\mathbb{N}}$ *be an orthonormal basis of the selfadjoint compact operator* $F_\sharp := |\operatorname{Re}(\gamma^{-1}F)| + \operatorname{Im}(\gamma^{-1}F)$. *A point* $y \in \mathbb{R}^2$ *belongs to* D *if and only if*

$$\sum_{j=1}^{\infty} \frac{\left|\langle \Phi_\infty(\cdot, y), \psi_j\rangle_{L^2(\mathbb{S}^1)}\right|^2}{\lambda_j} < \infty. \tag{I.27}$$

We emphasize that the definition of $F_\sharp$ is in this entire thesis of local nature. Later on in Theorem I.7 we will for instance omit the constant γ^{-1} in the definition of $F_\sharp$. The precise definition of $F_\sharp$ will always be stated explicitely.

Proof. The claim follows directly from an application of Theorem I.7 to the factorization $\gamma^{-1}F = H^*TH$. All the assumptions of Theorem I.7 have been checked in Lemma I.5. Note that (I.25) implies that $\operatorname{Im} T = (T - T^*)/(2\mathrm{i})$ is positive semidefinite, since

$$\operatorname{Im}\left(\int_D \overline{f}\, Tf \,\mathrm{d}x\right) = \int_D \overline{f}\,(\operatorname{Im} T) f \,\mathrm{d}x.$$

Therefore Theorem I.7 implies that $\operatorname{Range}(F_\sharp^{1/2}) = \operatorname{Range}(G)$. Since (λ_j, ψ_j) is an eigensystem of $F_\sharp$ we observe that $(\lambda_j^{1/2}, \psi_j)$ is an eigensystem of $F_\sharp^{1/2}$ and Picard's theorem [46] implies that

$$f \in L^2(\mathbb{S}^1) \text{ belongs to } \operatorname{Range}(G) \quad \Longleftrightarrow \quad \sum_{j=1}^{\infty} \frac{|\langle f, \psi_j \rangle|^2}{\lambda_j} < \infty.$$

Part (c) of Theorem I.5 now yields the claim of the theorem. □

Recall from (I.15) that the far field $\Phi_\infty(\cdot, y)$ is given by $\hat{x} \mapsto \gamma \exp(-\mathrm{i}k\, \hat{x} \cdot y)$. Also note that the series in (I.27) is constantly referred to as a Picard series in the following, due to its occurrence in Picard's theorem, where this series tests the range of a compact operator. In a numerical implementation of the Factorization method, one picks grid points z_i, $i = 1, \dots, I$, in a certain test domain and computes the inverse of the value of the series in (I.27) for each z_i. Of course, in numerical computations we cannot compute the entire series but only compute a finite approximation with summands that are afflicted with certain errors. Nevertheless, one might hope that this procedures leads to small values at points z outside the support D of the contrast q and to large values at points inside D. We will rigorously study the behavior of the series in (I.27) for perturbed data in Section I-5.

I-4. A Range Identity for Factorizations

In this section we prove the fundamental functional analytic result for the Factorization method, which is a range identity for factorizations. In the proof of this result, already announced before Theorem I.4, we follow the approach of Grinberg and Kirsch [49, 47] and consider the auxiliary operator $F_\sharp$, which is easily computable from the far field operator F. However, we modify the assumptions of the theorem such that they fit to the situation of this (and also of the following) chapters, where the non-selfadjoint part of the middle operator T is non-negative, but fails to be positive. More precisely, $\operatorname{Im} \int_D \overline{f}\, Tf \, \mathrm{d}x \geq 0$ is valid for all $f \in L^2(D)$, but no strict inequality holds. The new feature is that assumptions of either injectivity of the corresponding real part $\operatorname{Re} T$ (which is sometimes also called the selfadjoint part of F) or else positivity of the imaginary part $\operatorname{Im} T$ (sometimes also called the non-selfadjoint part of F) appear in a kind of "mixed" form. Concerning the characterization result of this chapter in Theorem I.4, we remark that our approach has the advantage that the consideration of the transmission eigenvalues of the direct scattering problem can be completely avoided, compare [49, Chapter 4].

Before stating the theorem, we recall the definitions of real and imaginary part of an operator T on a Hilbert space, $\operatorname{Re} T := 1/2\,(T + T^*)$ and $\operatorname{Im} T := 1/(2\mathrm{i})\,(T - T^*)$.

THEOREM I.7. *Let* $\mathsf{X} \subset \mathsf{U} \subset \mathsf{X}^*$ *be a Gelfand triple with Hilbert space* U *and reflexive Banach space* X *such that the embedding is dense. Furthermore, let* V *be a second Hilbert space and* $F : \mathsf{V} \to \mathsf{V}$, $H : \mathsf{V} \to \mathsf{X}$ *and* $T : \mathsf{X} \to \mathsf{X}^*$ *be linear and bounded operators with*

$$F = H^* T H.$$

We make the following assumptions:

(a) H *is compact and injective.*

(b) $\operatorname{Re} T$ *has the form* $\operatorname{Re} T = T_0 + T_1$ *with some coercive operator* T_0 *and some compact operator* $T_1 : \mathsf{X} \to \mathsf{X}^*$.

(c) $\operatorname{Im} T$ *is non-negative on* X, *i.e.,* $\langle \operatorname{Im} T\phi, \phi\rangle \geq 0$ *for all* $\phi \in \mathsf{X}$.

Moreover, we assume that one of the following conditions is fulfilled.

(d) T *is injective.*

(e) $\operatorname{Im} T$ *is positive on the finite dimensional null space of* $\operatorname{Re} T$, *i.e., for all* $\phi \neq 0$ *such that* $\operatorname{Re} T\phi = 0$ *it holds* $\langle \operatorname{Im} T\phi, \phi\rangle > 0$.

Then the operator $F_\sharp := |\operatorname{Re} F| + \operatorname{Im} F$ *is positive definite and the ranges of* $H^* : \mathsf{X}^* \to \mathsf{V}$ *and* $F_\sharp^{1/2} : \mathsf{V} \to \mathsf{V}$ *coincide.*

Proof. We first recall from [47] that it is sufficient to assume that $\mathsf{X} = \mathsf{U}$ is a Hilbert space and that H has dense range in U. The reduction to the Hilbert space case follows from the introduction of the coercive square root $T_0^{1/2} : \mathsf{X} \to \mathsf{U}$, see, e.g., [76, Theorem 12.33], since

$$F = H^* T H = \left(H^* T_0^{1/2}\right)\left(T_0^{-1/2} T T_0^{1/2}\right)\left(T_0^{-1/2} H\right) =: \widetilde{H}^* \widetilde{T} \widetilde{H}.$$

If the range of $\widetilde{H}$ is not dense in U, we replace U by its closed subspace $\overline{\text{Range}(H)}$ using the orthogonal projector P from U onto $\overline{\text{Range}(H)}$. Since $PH = H$, the factorization $F = H^*PTPH$ holds and all properties required in the claim of the theorem are preserved. Hence, we can assume that X is a Hilbert space, which is in the remainder of the proof denoted as U, and that H has dense range.

The factorization of F implies that $\operatorname{Re} F$ is compact and selfadjoint. By the spectral theorem for such operators, there exists a complete orthonormal eigensystem $(\lambda_j, \psi_j)_{j\in\mathbb{N}}$ of $\operatorname{Re} F$. In consequence, the spaces

$$\mathsf{V}^+ = \operatorname{span}\{\psi_j : \lambda_j > 0\} \qquad \text{and} \qquad \mathsf{V}^- = \operatorname{span}\{\psi_j : \lambda_j \leq 0\}$$

are invariant under $\operatorname{Re} F$ and satisfy $\mathsf{V} = \mathsf{V}^+ \oplus \mathsf{V}^-$. We set $\mathsf{U}^- = H\mathsf{V}^-$.

In the next step we show that U^- is finite dimensional: The selfadjoint part $\operatorname{Re} T$ of the middle operator $T : \mathsf{U} \to \mathsf{U}$ is by assumption sum of a coercive and a compact operator and we denote by $(\mu_j, \phi_j)_{j\in\mathbb{N}}$ an eigensystem of $\operatorname{Re} T$. We set $\mathsf{W}^\pm = \operatorname{span}\{\phi_j : \mu_j \gtrless 0\}$ and note that W^- is finite dimensional. Let now $\phi = H\psi \in \mathsf{U}^-$ with (unique) decomposition $\phi = \phi^+ + \phi^-$, $\phi^\pm \in \mathsf{W}^\pm$. Since $\psi \in \mathsf{V}^-$, we note that

$$\begin{aligned} 0 \geq \langle \operatorname{Re} F\psi, \psi\rangle &= \langle (\operatorname{Re} T)H\psi, H\psi\rangle = \langle \operatorname{Re} T(\phi^+ + \phi^-), \phi^+ + \phi^-\rangle \\ &= \langle \operatorname{Re} T\phi^+, \phi^+\rangle + \langle \operatorname{Re} T\phi^-, \phi^-\rangle \geq c\|\phi^+\|^2 - \|\operatorname{Re} T\|\|\phi^-\|^2, \end{aligned}$$

thus, $\|\phi\|^2 = \|\phi^+\|^2 + \|\phi^-\|^2 \leq C\|\phi^-\|^2$. This shows that the mapping $\phi \mapsto \phi^-$ is boundedly invertible from U^- into W^-. Consequently, U^- is finite dimensional, because W^- is finite dimensional.

The denseness of the range of H implies that the sum $\overline{H\mathsf{V}^+} + \mathsf{U}^-$ is dense in U. Since U^- is a finite dimensional and therefore complemented subspace, compare [81], we can choose a closed subspace U^+ of $\overline{H\mathsf{V}^+}$ such that the (non-orthogonal) sum $\mathsf{U} = \mathsf{U}^+ \oplus \mathsf{U}^-$ is direct. Let moreover $U^o := \overline{H\mathsf{V}^+} \cap \mathsf{U}^-$ be the intersection of $\overline{H\mathsf{V}^+}$ and U^-. We denote by $P_{\mathsf{U}^\pm} : \mathsf{U} \to \mathsf{U}^\pm$ the canonical bounded projections, that is, every $\phi \in \mathsf{U}$ has the unique decomposition

$$\phi = P_{\mathsf{U}^+}\phi + P_{\mathsf{U}^-}\phi.$$

Both operators $P_{\mathsf{U}^\pm}$ are bounded and $P_{\mathsf{U}^+} - P_{\mathsf{U}^-}$ is an isomorphism, since

$$(P_{\mathsf{U}^+} - P_{\mathsf{U}^-})^2 = P_{\mathsf{U}^+}^2 + P_{\mathsf{U}^-}^2 - P_{\mathsf{U}^+}P_{\mathsf{U}^-} - P_{\mathsf{U}^-}P_{\mathsf{U}^+} = P_{\mathsf{U}^+} + P_{\mathsf{U}^-} = \operatorname{Id}.$$

Unfortunately, $\mathrm{Re}\,T$ fails to keep the spaces $\mathsf{U}^{\pm}$ invariant. However, the action of $\mathrm{Re}\,T$ can be seen as "flipping the spaces $\mathsf{U}^{\pm}$", as the following argument shows. From the factorization

$$\mathrm{Re}\,F = H^* \,(\mathrm{Re}\,T)\, H$$

and the definition of $\mathsf{U}^{\pm}$ we obtain $H^* \,\mathrm{Re}\,T(\mathsf{U}^-) = \mathrm{Re}\,F(\mathsf{V}^-) \subset \mathsf{V}^-$. Note also that by definition $\mathsf{U}^+ \subset \overline{H(\mathsf{V}^+)}$. In consequence, for $\phi^- \in \mathsf{U}^-$ and $\psi^+ \in \mathsf{V}^+$ we have

$$0 = \langle H^*(\mathrm{Re}\,T)\phi^-, \psi^+\rangle = \langle \mathrm{Re}\,T\phi^-, H\psi^+\rangle = \langle \phi^-, (\mathrm{Re}\,T)H\psi^+\rangle.$$

We conclude that $(\mathrm{Re}\,T)\mathsf{U}^- \subset (H(\mathsf{V}^+))^{\perp} = (\mathsf{U}^+ \oplus \mathsf{U}^o)^{\perp} \subset (\mathsf{U}^+)^{\perp}$ and, moreover, $(\mathrm{Re}\,T)\mathsf{U}^+ \subset (\mathsf{U}^-)^{\perp}$. Indeed, for $\phi^+ \in \mathsf{U}^+$ there is a sequence $\psi_n \in \mathsf{V}$ such that $H\psi_n \to \phi$ and $H\psi_n \in (\mathsf{U}^-)^{\perp}$, thus, $\phi^+ \in (\mathsf{U}^-)^{\perp}$.

Now we prove that $\mathsf{U}^o := \overline{H\mathsf{V}^+} \cap \mathsf{U}^-$ is contained in the kernel of $\mathrm{Re}\,T$. This allows later on to show a factorization for $F_\sharp$. For $\phi^o \in \overline{H\mathsf{V}^+} \cap \mathsf{U}^-$ we observe as above that $\langle \mathrm{Re}\,T\phi^o, \phi^o\rangle = \langle \mathrm{Re}\,TH\psi, H\psi\rangle = \langle \mathrm{Re}\,F\psi, \psi\rangle \leq 0$ since $\phi^o \in \mathsf{U}^-$. Since ϕ^o also belongs to $\overline{H\mathsf{V}^+}$, a density argument implies $\langle \mathrm{Re}\,T\phi^o, \phi^o\rangle \geq 0$, that is, $\langle \mathrm{Re}\,T\phi^o, \phi^o\rangle = 0$. However, by the mapping properties of $\mathrm{Re}\,T$ stated above, $\mathrm{Re}\,T\phi^o$ is orthogonal both to U^- and U^+ and hence $\langle \mathrm{Re}\,T\phi^o, \phi\rangle = 0$ for all $\phi \in \mathsf{U}$, implying that $\mathrm{Re}\,T\phi^o = 0$.

The result from the last paragraph allows now to prove a factorization of $F_\sharp$. Let $\psi \in \mathsf{V}$ and $\psi^{\pm}$ be its orthogonal projection on $\mathsf{V}^{\pm}$. Then

$$\begin{aligned}
|\mathrm{Re}\,F|\psi &= H^* \,(\mathrm{Re}\,T)\, H(\psi^+ - \psi^-) \\
&= H^* \mathrm{Re}\,T\big(P_{\mathsf{U}^+}H\psi^+ + P_{\mathsf{U}^-}H\psi^+ - P_{\mathsf{U}^+}H\psi^- - P_{\mathsf{U}^-}H\psi^-\big) \\
&= H^* \mathrm{Re}\,T(P_{\mathsf{U}^+}H\psi + 2 \underbrace{P_{\mathsf{U}^-}H\psi^+}_{\in \mathsf{U}^o \subset \ker(\mathrm{Re}\,T)} - P_{\mathsf{U}^-}H\psi) \\
&= H^* \mathrm{Re}\,T(P_{\mathsf{U}^+} - P_{\mathsf{U}^-})H\psi = H^*|\mathrm{Re}\,T|H\psi.
\end{aligned}$$

This factorization of $|\mathrm{Re}\,F|$ yields a factorization of $F_\sharp$,

$$F_\sharp = |\mathrm{Re}\,F| + \mathrm{Im}\,F = H^*|\mathrm{Re}\,T|H + H^*\mathrm{Im}\,TH = H^*T_\sharp H,$$

where $T_\sharp = |\mathrm{Re}\,T| + \mathrm{Im}\,T = \mathrm{Re}\,T(P_{\mathsf{U}^+} - P_{\mathsf{U}^-}) + \mathrm{Im}\,T$. Moreover, due to

$$\langle \mathrm{Re}\,T(P_{\mathsf{U}^+} - P_{\mathsf{U}^-})H\phi, H\phi\rangle = \langle |\mathrm{Re}\,F|\phi, \phi\rangle \geq 0, \qquad \phi \in \mathsf{V},$$

and the fact that the range of H is dense in U, we conclude that $|\mathrm{Re}\, T|$ is non-negative on U. The following general inequality for an arbitrary bounded non-negative operator A on a Hilbert space H has been shown in [47, Equation 4.5]:

$$\langle A\psi, \psi \rangle \geq \frac{1}{\|A\|} \|A\psi\|^2, \qquad \psi \in \mathsf{H}. \tag{I.28}$$

Since $T_\sharp$ is a non-negative operator (as sum of two non-negative operators) we have

$$\langle T_\sharp \psi, \psi \rangle \geq \frac{1}{\|T_\sharp\|} \|T_\sharp \psi\|^2, \qquad \psi \in \mathsf{U}. \tag{I.29}$$

We finish the proof using one of the assumptions (d) or (e).

(d) First, we show that assumption (d) implies assumption (e). Under assumption (d), let ϕ belong to the null space of $\mathrm{Re}\, T$ and suppose that $\langle \mathrm{Im}\, T\phi, \phi \rangle = 0$. We need to show that this implies that $\phi = 0$. By definition of the real part of an operator,

$$T\phi + T^*\phi = 0. \tag{I.30}$$

Furthermore, (I.28) applied to $\mathrm{Im}\, T$ states that

$$0 = \langle \mathrm{Im}\, T\phi, \phi \rangle \geq \frac{1}{\|\mathrm{Im}\, T\|} \|\mathrm{Im}\, T\phi\|^2,$$

and hence $\|\mathrm{Im}\, T\phi\| = 0$ and $\mathrm{Im}\, T\phi = 0$. By definition of the imaginary part, this is to say that $T\phi - T^*\phi = 0$. Combining this equation with (I.30) yields that $T\phi = 0$, and assumption (d), which supposes the injectivity of T, implies that $\phi = 0$. We have hence proven that $\langle \mathrm{Im}\, T\phi, \phi \rangle > 0$ for all $0 \neq \phi \in \ker(\mathrm{Re}\, T)$. This is precisely assumption (e), which is considered next.

(e) Assuming assumption (e), we show that $T_\sharp$ is injective. Suppose that $T_\sharp \phi = 0$. Recall from above that $T_\sharp$ is the sum of two non-negative operators. Since

$$0 = \langle |\mathrm{Re}\, T|\phi, \phi \rangle + \langle \mathrm{Im}\, T\phi, \phi \rangle, \tag{I.31}$$

we observe that both terms on the right hand side have to vanish. Exploiting again (I.28), we note that $\langle |\mathrm{Re}\, T|\phi, \phi \rangle = 0$ implies that $|\mathrm{Re}\, T|\phi = 0$. Moreover, due to selfadjointness,

$$|\mathrm{Re}\, T| = (\mathrm{Re}\, T)(P_{\mathsf{U}^+} - P_{\mathsf{U}^-}) = (P_{\mathsf{U}^+} - P_{\mathsf{U}^-})^* \, \mathrm{Re}\, T$$

and $(P_{\mathsf{U}^+} - P_{\mathsf{U}^-})^*$ is an isomorphism (since $P_{\mathsf{U}^+} - P_{\mathsf{U}^-}$ is an isomorphism). Consequently, $\operatorname{Re} T\phi = 0$. Assumption (e) now implies that $\langle \operatorname{Im} T\phi, \phi\rangle > 0$ if $\phi \neq 0$. However, we showed above that $\langle \operatorname{Im} T\phi, \phi\rangle = 0$, that is, $\phi = 0$ and therefore $T_\sharp$ is injective.

Hence, by assumption (d) or (e), $T_\sharp$ is an injective Fredholm operator of index 0 (Fredholmness is due to assumption (b) of this theorem) and hence boundedly invertible. By (I.29) we obtain

$$\langle T_\sharp \psi, \psi\rangle \geq \frac{1}{\|T_\sharp\|}\|T_\sharp \psi\|^2 \geq C\|\psi\|^2 \qquad \text{for all } \psi \in \mathsf{U}.$$

Now, as $T_\sharp$ has been shown to be coercive, we can conclude as in [47]: The square root $T_\sharp^{1/2}$ of $T_\sharp$ is also coercive on U, see, e.g., [76, Theorems 12.32, 12.33], hence the inverse $T_\sharp^{-1/2}$ is bounded and we can write

$$F_\sharp = F_\sharp^{1/2}\left(F_\sharp^{1/2}\right)^* = H^* T_\sharp H = \left(H^* T_\sharp^{1/2}\right)\left(H^* T_\sharp^{1/2}\right)^*.$$

However, if two positive operators agree, then the ranges of their square roots agree, as the next lemma shows.

Lemma I.8. *Let V, U_1 and U_2 be Hilbert spaces and $A_j : \mathsf{U}_j \to \mathsf{V}$, $j = 1, 2$, bounded and injective such that $A_1 A_1^* = A_2 A_2^*$. Then the ranges of A_1 and A_2 coincide and $A_1^{-1} A_2$ is an isometric isomorphism from U_2 onto U_1.*

We refer to [47, Lemma 2.4] for a proof. Setting $A_1 = F_\sharp^{1/2}$ and $A_2 = H^* T_\sharp^{1/2}$, the last lemma states that the ranges of $F_\sharp^{1/2}$ and $H^* T_\sharp^{1/2}$ agree and that $F_\sharp^{-1/2} H^* T_\sharp^{1/2}$ is an isomorphism from U to V. Since $T_\sharp^{1/2}$ is an isomorphism on U, we conclude that the range of $H^* T_\sharp^{1/2}$ equals the range of H^* and that $F_\sharp^{-1/2} H^* : \mathsf{U} \to \mathsf{V}$ is bounded with bounded inverse. The proof of the theorem is complete. □

I-5. Regularization of the Factorization Method: Preliminaries

The Factorization method is an elegant tool to solve the inverse medium scattering problem, since it provides an explicit representation of the support D of the contrast q by the Picard series. Recall from (I.27) that

$$y \in D \quad \Longleftrightarrow \quad \sum_{j=1}^{\infty} \frac{|\langle \Phi_\infty(\cdot, y), \psi_j \rangle|^2}{\lambda_j} < \infty, \tag{I.32}$$

where $\Phi_\infty(\cdot, y)$ is the far field pattern of a point source at $y \in \mathbb{R}^2$ and $(\lambda_j, \psi_j)_{j\in\mathbb{N}}$ denotes an eigensystem of the selfadjoint positive operator $F_\sharp$. Equivalently, the sequence

$$N \mapsto \sum_{j=1}^{N} \frac{|\langle \Phi_\infty(\cdot, y), \psi_j \rangle|^2}{\lambda_j} \quad \text{is bounded} \quad \Longleftrightarrow \quad y \in D.$$

In this section, we are going to investigate what happens to the criterion in (I.32) when we do not know the far field operator F exactly but only approximately. One might for instance have in mind the case of a finite dimensional measured approximation of F. For example, assume we are given

$$u^s_\infty(\theta_i, \theta_j) \text{ for } i, j = 0, \dots 2n,\ \theta_j := (\sin(\pi j/n), \cos(\pi j/n)), \tag{I.33}$$

for some $n \in \mathbb{N}$. Given the function values $f(\pi j/n)$, $j = 0, \dots, 2n$, of a real analytic and 2π periodic real function f, its trigonometric interpolation polynomial, see [51], is given by

$$P_n(f)(x) := \sum_{j=0}^{n} a_j \cos(jx) + \sum_{j=1}^{n-1} b_j \sin(jx)$$

with coefficients

$$a_0 = \frac{1}{2n} \sum_{m=0}^{2n-1} f(\pi m/n), \qquad a_n = \frac{1}{2n} \sum_{m=0}^{2n-1} (-1)^m f(\pi m/n),$$

$$a_j + \mathrm{i}\, b_j = \frac{1}{n} \sum_{m=0}^{2n-1} f(\pi m/n) e^{\mathrm{i}\pi j m/n}, \qquad j = 1, \dots, n-1.$$

The trigonometric polynomial $P_n(f)$ is uniquely determined by the interpolation property

$$P_n(f)(\pi j/n) = f(\pi j/n), \qquad j = 0, \dots, 2n.$$

For the far field pattern u_∞^s, which depends on two variables, the product interpolation polynomial $P_{n,n}(u_\infty^s)$ is found by first applying P_n to the first variable of u_∞^s (the second variable being though of as a parameter). The result is a function of two variables, which is for fixed second argument a trigonometric polynomial in the first variable, say,

$$P_n\big(u_\infty^s(\cdot, x_2)\big)(x_1) = \sum_{j=0}^{n} a_j(x_2)\cos(jx_1) + \sum_{j=1}^{n-1} b_j(x_2)\sin(jx_1).$$

Afterwards one applies P_n to the second variable of this function of two variables. Thereby one obtains a function $P_{n,n}(u_\infty^s)$ periodic in two variables which is a trigonometric polynomial in each of its two arguments. Moreover, exploiting the interpolation property of P_n, we have that

$$P_{n,n}(u_\infty^s)(\pi i/n,\, \pi j/n) = u_\infty^s(\pi i/n,\, \pi j/n), \qquad i,j = 0,\dots,2n.$$

Using the exponential convergence $\|P_n(f) - f\|_\infty = \mathcal{O}(e^{-ns})$ for some $s > 0$ proved in [51] and arguments from [52, Section 11.2] one can show that the "interpolated" approximation of the far field operator

$$F_n\psi = \int_{\mathbb{S}^1} \psi(\theta) P_{n,n}\left(u_\infty^s\right)(\cdot,\theta)\,\mathrm{d}s(\theta), \qquad \psi \in L^2(\mathbb{S}^1),$$

converges exponentially to F in the operator norm, $\|F - F_n\| \le \mathcal{O}(e^{-ns})$. Obviously, F_n is a finite dimensional operator and a characterization result as Theorem I.6 cannot hold for F_n, since the "approximate" eigenvalues $\lambda_{j,n}$ of $F_\sharp^n := |\operatorname{Re} F_n| + \operatorname{Im} F_n$ vanish for j large enough. However, one might wonder whether Picard's criterion (I.32) holds for F_n in some appropriate limit sense as $n \to \infty$. This will indeed be the case, if one applies a suitable regularization by spectral cut-off.

Let us also note here the study in [26] where the good performance of a sampling method for low frequency is explained by the validity of the Factorization method in the low frequency limit. This way of explanation is in the same spirit as the asymptotic study of the Factorization method in the rest of this chapter. Let us also note another (but different) asymptotic factorization technique developed in [30] using asymptotic expansions for small scattering objects.

Regularization of the Picard criterion (I.32) is easily motivated to be necessary to obtain meaningful reconstructions. Since F is compact, its eigenvalues λ_j tend to zero and in (I.32) one divides by λ_j. Hence, numerical errors in λ_j are strongly amplified: If the eigenvalues λ_j of $F_\sharp$ are slightly perturbed, say, by using the $\lambda_{j,n}$ of $F_\sharp^n$, the value of the truncated Picard series for some fixed truncation index $n \in \mathbb{N}$ is likely to become completely useless. Here enters the ill-posedness of the inverse medium problem, see, e.g., [22]. In numerical experiments, the Picard criterion in (I.32) has nevertheless been shown to yield good reconstructions of scattering objects. We are going to explain this behavior at least partly using perturbation theory and a a suitable regularization of (I.32) by spectral cut-off. The technique we use here has been worked out in the paper [57] and applied to the so-called complete electrode model in impedance tomography in [58].

Before we treat the Factorization method with perturbed data we need to have a look at some basics of perturbation theory for compact normal operators, which can be found, e.g., in [41]. It is well known [76, Theorem 12.29] that such an operator $F : \mathsf{H} \to \mathsf{H}$ on a Hilbert space H has an eigensystem, which we denote by $(\lambda_j, \psi_j)_{j\in\mathbb{N}}$,

$$F\psi = \sum_{j\in\mathbb{N}} \lambda_j \langle \psi, \psi_j \rangle \psi_j, \qquad \psi \in \mathsf{H}.$$

Here, $\langle \cdot, \cdot \rangle$ denotes the inner product in H which induces the norm $\| \cdot \|^2 = \| \cdot \|_{\mathsf{H}}^2$ on H. The spectrum of F is by definition the set of all values $\lambda \in \mathbb{C}$ such that $\lambda - F$ is not boundedly invertible. It is well known that the spectrum of a compact operator is the union of its eigenvalues and 0,

$$\sigma(F) = \bigcup_{j\in\mathbb{N}} \lambda_j \ \cup \{0\}.$$

The complement of $\sigma(F)$ in $\mathbb{C}$ is called the resolvent set and in this (open) set we can define the resolvent

$$R(\xi, F) = (\xi - F)^{-1} = \sum_{j\in\mathbb{N}} \frac{1}{\xi - \lambda_j} \langle \cdot, \psi_j \rangle \psi_j, \qquad \xi \in \mathbb{C} \setminus \sigma(F).$$

In general, a vector valued function $f : \Omega \subset \mathbb{C} \to \mathsf{X}$ is called holomorphic, if for each $z_0 \in \Omega$ the limit

$$\lim_{z\to z_0} \frac{f(z) - f(z_0)}{z - z_0} =: f'(z_0) \tag{I.34}$$

exists in the Banach space X. Note that Moreira's theorem holds for vector-valued holomorphic functions: If f is continuous in z and the integral $\int_\Delta f(z)\,\mathrm{d}z$ over the boundary of any closed triangle $\Delta \subset \Omega$ vanishes, then f is holomorphic. For details on complex variables in infinite dimensional spaces we refer to [67].

The resolvent is a holomorphic operator-valued function in $\mathbb{C} \setminus \sigma(F)$ and it is not difficult to show the following two crucial norm estimates

$$\begin{aligned} \|R(\xi, F)\|_{\mathsf{H}\to\mathsf{H}} &:= \sup_{0\neq\psi\in\mathsf{H}} \frac{\|R(\xi, F)\psi\|_{\mathsf{H}}}{\|\psi\|_{\mathsf{H}}} = \sup_{j\in\mathbb{N}} \left(|\xi - \lambda_j|^{-1}\right), \\ \|FR(\xi, F)\|_{\mathsf{H}\to\mathsf{H}} &= \sup_{j\in\mathbb{N}} \left(|\lambda_j|\, |\xi - \lambda_j|^{-1}\right). \end{aligned} \tag{I.35}$$

These estimates allow to prove two perturbation lemmas in the sequel. Note that we suppress from now on the subscripts H and $\mathsf{H} \to \mathsf{H}$ in the norms and operator norms, respectively.

LEMMA I.9 (Perturbation Lemma I). *Let F_1, $F_2 : \mathsf{H} \to \mathsf{H}$ be compact normal operators. Then* $\operatorname{dist}(\sigma(F_1), \sigma(F_2)) \leq \|F_1 - F_2\|$, *that is,*

$$\sup_{\xi\in\sigma(F_1)} \operatorname{dist}(\xi, \sigma(F_2)) \leq \|F_1 - F_2\| \qquad \textit{and vice versa.}$$

Proof. It is by symmetry sufficient to prove that $\operatorname{dist}(\xi, \sigma(F_2)) > \|F_1 - F_2\|$ implies $\xi \in \mathbb{C} \setminus \sigma(F_1)$. Hence, assume that $\operatorname{dist}(\xi, \sigma(F_2)) > \|F_1 - F_2\|$ and that $F_1 \neq F_2$, to avoid trivial cases. From (I.35) we conclude that $\|R(\xi, F_2)\| < \|F_1 - F_2\|^{-1}$, hence, by a Neumann series argument, $\operatorname{Id} - (F_1 - F_2)R(\xi, F_2)$ is invertible. The resolvent identity

$$R(\xi, F_1) - R(\xi, F_2) = R(\xi, F_1)(F_1 - F_2)R(\xi, F_2) \tag{I.36}$$

for $\xi \notin \sigma(F_1) \cup \sigma(F_2)$, which can be proved by multiplication with $\xi - F_1$ and $\xi - F_2$ from left and right, respectively, shows that

$$R(\xi, F_1) = R(\xi, F_2)^{-1} \left[\operatorname{Id} - (F_1 - F_2)R(\xi, F_2)\right]^{-1}.$$

Therefore $R(\xi, F_1)$ is a bounded operator and thus $\xi \in \mathbb{C} \setminus \sigma(F_1)$, which was to show. □

A similar lemma theorem holds for the eigenspaces instead of the eigenvectors of a compact normal operator. To state such a result, it is convenient to use the concept of a *spectral projection*. For a compact normal operator one can define the spectral projection on the eigenspace associated with some eigenvalue λ_j by

$$P_j(F)\psi := \sum_{n:\,\lambda_n=\lambda_j} \langle \psi, \psi_n \rangle \psi_n, \qquad \psi \in \mathsf{H}. \tag{I.37}$$

One can express P_j in a more suitable way for our later purpose using complex function theory. Recall that the index of point z with respect to a closed smooth and positively oriented path in the complex plane γ is defined as

$$\mathrm{ind}_\gamma(z) = \frac{1}{2\pi \mathrm{i}} \int_\gamma \frac{\mathrm{d}\xi}{\xi - z}, \qquad z \in \mathbb{R}^2 \setminus \gamma.$$

The function ind_γ is integer-valued and vanishes in the unbounded component of $\mathbb{R}^2 \setminus \gamma$, see, e.g., [75, Theorem 10.10].

PROPOSITION I.10. *Let $F : \mathsf{H} \to \mathsf{H}$ be compact and normal with eigensystem $(\lambda_j, \psi_j)_{j\in\mathbb{N}}$ and fix $n \in \mathbb{N}$. Let γ be a closed smooth and positively oriented path in the complex plane such that γ separates λ_n from the rest of the spectrum of F: $\mathrm{ind}_\gamma(\lambda_j) = \delta_{\lambda_n,\lambda_j}$, $j \in \mathbb{N}$. Then*

$$P_n(F)\psi = \frac{1}{2\pi \mathrm{i}} \int_\gamma R(\xi, F)\psi \,\mathrm{d}\xi, \qquad \psi \in \mathsf{H}. \tag{I.38}$$

For a proof using Cauchy's integral formula we refer to [57, Proposition 4.3]. The integral in the last formula is a vector-valued integral with Hilbert space values. For details on integration in Banach spaces we refer to [76, Chapter 3]. Note that (I.38) has the advantage that we can replace F by any operator which has no spectrum on the curve γ such that the resolvent is well defined. Therefore we denote the spectral projection sometimes by

$$P_\gamma(F) = \frac{1}{2\pi \mathrm{i}} \int_\gamma R(\xi, F)\,\mathrm{d}\xi,$$

to emphasize the dependence on the path γ. Using the concept of the spectral projection we can now state the second perturbation lemma, which deals with perturbations of eigenspaces.

LEMMA I.11 (Perturbation Lemma II). *Let F_1, $F_2 : \mathsf{H} \to \mathsf{H}$ be compact normal operators. Consider a closed smooth positively oriented curve γ in the complex plane such that γ separates the eigenvalue λ_n of F_1 from* *the rest of the spectrum of* F_1. *The associated projection* $P_\gamma(F_1)$ *gives rise to an orthogonal decomposition* $\mathsf{H} = M_1 \oplus M_1^\perp$, *where* $P_\gamma(F_1)$ *is the orthogonal projection onto* M_1. *If* $\|F_1 - F_2\| < \operatorname{dist}(\lambda_n, \gamma)$ *then* γ *separates the spectrum of* F_2 *into two parts, too. The analogous decomposition* $H = M_2 \oplus M_2^\perp$, *defined with the help of* $P_\gamma(F_2)$ *(the orthogonal projection onto* M_2*), has the property that* M_1 *is norm isomorphic to* M_2. *Moreover, if* $\|F_1 - F_2\| \leq \delta$ *and* $\operatorname{dist}(\lambda_n, \sigma(F_1) \setminus \{\lambda_n\}) \geq 2\rho$ *for some* $\rho > \delta > 0$, *then*

$$\|P_\gamma(F_1) - P_\gamma(F_2)\| \leq \frac{\delta}{\rho - \delta}. \tag{I.39}$$

Proof. Except for estimate (I.39) a proof of the theorem can be found in [41, Chapter IV, Theorem 3.16, page 212]. Here we merely prove the given estimate following [57, Theorem 4.2]. Note also that orthogonality of the projection $P_\gamma(F_j)$, $j = 1, 2$, follows from (I.37). From Proposition I.10 we obtain that the projections $P_\gamma(F_1)$ and $P_\gamma(F_2)$ do not change if we redefine γ to be the circle of radius ρ around λ_n. Indeed, the eigenvalues of F_1 and F_2 inside and outside the curve γ are precisely the same after this redefinition as before.

The definition of the spectral projection in (I.38) and a simple estimate yield

$$\|P_\gamma(F_1) - P_\gamma(F_2)\| \leq \frac{|\gamma|}{2\pi} \sup_{\xi \in \gamma} \|R(\xi, F_1) - R(\xi, F_2)\|,$$

where $|\gamma|$ denotes the length of the curve γ. Obviously, the length of γ is now $2\pi\rho$. The resolvent identity (I.36) and (I.35) imply that

$$\begin{aligned}\sup_{\xi \in \gamma} \|R(\xi, F_1) - R(\xi, F_2)\| &= \sup_{\xi \in \gamma} \|R(\xi, F_1)(F_1 - F_2) R(\xi, F_2)\| \\ &\leq \delta \sup_{\xi \in \gamma} \left(\sup_{\lambda \in \sigma(F_1)} \left(|\lambda - \xi|^{-1}\right) \sup_{\mu \in \sigma(F_2)} \left(|\mu - \xi|^{-1}\right) \right).\end{aligned}$$

The first supremum inside the bracket of the last equation can be bounded by $1/\rho$, since ξ is a point on the curve γ and λ belongs to the spectrum of F_1. Indeed, the distance of ξ to λ_n equals ρ and the distance of ξ to $\sigma(F_1) \setminus \{\lambda_n\}$ is grater or equal than ρ, using our assumption $\operatorname{dist}(\lambda_n, \sigma(F_1) \setminus \{\lambda_n\}) \geq 2\rho$ and the (inverse) triangle inequality. The second supremum can be bounded by $1/(\rho - \delta)$ using the bound $\|F_1 - F_2\| \leq \delta$ and another (inverse) triangle inequality. □

I-6. A Perturbation Lemma for the Factorization Method

From now on, we make the general assumption that $F_\sharp : \mathsf{H} \to \mathsf{H}$ is a compact positive selfadjoint operator on some Hilbert space H with eigensystem $(\lambda_j, \psi_j)_{j\in\mathbb{N}}$. Then Picard's criterion characterizes the range of $F_\sharp^{1/2}$: For $\phi \in \mathsf{H}$ the sequence

$$n \mapsto \sum_{j=1}^{n} \frac{|\langle \phi, \psi_j \rangle|^2}{\lambda_j} \text{ is bounded} \iff \phi \in \mathrm{Range}(F_\sharp^{1/2}). \tag{I.40}$$

In this section, we study the behavior of the Picard criterion in case that the eigensystem (λ_j, ψ_j) is only known approximately. The Factorization method itself will merely play a secondary role in the remainder of this chapter and one can interpret the following study purely as an analysis of a regularization method for Picard's criterion (I.40).

However, for the specific example of the inverse medium problem, $F_\sharp$ denotes the auxiliary operator from Theorem I.6, $\mathsf{H} = L^2(\mathbb{S}^1)$ and we used testfunctions $\phi_y = \Phi_\infty(\cdot, y)$ to obtain a characterization of the domain D. Our analysis in this section is based on the fundamental assumption that we know some sequence of operators $F_\sharp^n$ which converges to $F_\sharp$ as $n \to \infty$. Unfortunately, in the inverse medium problem, we measure an approximation F_n of the far field operator F, from which we compute an approximation $F_\sharp^n := |\mathrm{Re}\, F_n| + \mathrm{Im}\, F_n$ of $F_\sharp := |\mathrm{Re}\, F| + \mathrm{Im}\, F$. Hence, for completeness, we still need to investigate whether $F_\sharp^n \to F_\sharp$ if $F_n \to F$ as $n \to \infty$. This is not obvious, since $F_\sharp$ depends nonlinearly on F. However, the following estimate which we found in [82], answers this question: For A, B bounded operators on a Hilbert space H and $p > 0$ there exists a constant C_p such that

$$\| |A|^p - |B|^p \| \leq C_p \big(1 + |\ln(\|A - B\|)|\big) \|A - B\|^{\min(1,p)}. \tag{I.41}$$

Using the latter estimate for $p = 1$ we conclude that

$$\begin{aligned} \|F_\sharp^n - F_\sharp\| &\leq \| |\mathrm{Re}\, F_n| - |\mathrm{Re}\, F| \| + \|\mathrm{Im}\, F_n - \mathrm{Im}\, F\| \\ &\leq \big(C_1 \big(1 + |\ln(\|F_n - F\|)|\big) + 1\big) \|F_n - F\|. \end{aligned}$$

Now we see that $\|F_\sharp^n - F_\sharp\| \to 0$ if $\|F_n - F\| \to 0$, with any rate $\alpha < 1$, that is, for every $\alpha \in (0,1)$ there is $C(\alpha)$ such that

$$\|F_\sharp^n - F_\sharp\| \le C_\alpha \|F_n - F\|^\alpha.$$

We set

$$d_n := \|F_\sharp^n - F_\sharp\|, \qquad n \in \mathbb{N}, \tag{I.42}$$

and denote by $(\lambda_{j,n}, \psi_{j,n})_{j\in\mathbb{N}}$ an eigensystem of $F_\sharp^n$ and by $(\lambda_j, \psi_j)_{j\in\mathbb{N}}$ an eigensystem of $F_\sharp$. Then $|\lambda_{j,n} - \lambda_j| \le d_n$ for all j and $n \in \mathbb{N}$ due to Lemma I.9.

Let us first motivate our method to regularize the Picard criterion (I.40) with perturbed data by a simple example, where we only perturb the eigenvalues λ_j of $F_\sharp$, but not the eigenvectors. Define the truncation index $\widetilde{T}_n$ by $\widetilde{T}_n := \max\{k \in \mathbb{N} : \lambda_{k,n} \ge (2+\gamma)d_n\}$, for $\gamma > 0$ a fixed parameter. Since $d_n \to 0$ but $\lambda_{k,n} \to \lambda_k$ as $n \to \infty$, the index $\widetilde{T}_n$ tends to infinity as $n \to \infty$. Moreover, due to $\lambda_j \ge \lambda_{j,n} - d_n \ge (1+\gamma)d_n$ we find

$$\begin{aligned}
\sum_{j=1}^{\widetilde{T}_n} \frac{|\langle \phi, \psi_j\rangle|^2}{\lambda_{j,n}} &\le \sum_{j=1}^{\widetilde{T}_n} \frac{|\langle \phi, \psi_j\rangle|^2}{\lambda_j - d_n} \\
&= \sum_{j=1}^{\widetilde{T}_n} \frac{|\langle \phi, \psi_j\rangle|^2}{\lambda_j} + \sum_{j=1}^{\widetilde{T}_n} \frac{|\langle \phi, \psi_j\rangle|^2 (\lambda_j - \lambda_j + d_n)}{\lambda_j(\lambda_j - d_n)} \\
&\le \sum_{j=1}^{\widetilde{T}_n} \frac{|\langle \phi, \psi_j\rangle|^2}{\lambda_j} + d_n \sum_{j=1}^{\widetilde{T}_n} \frac{|\langle \phi, \psi_j\rangle|^2}{\lambda_j(\gamma d_n)} \\
&\le \left(1 + \frac{1}{\gamma}\right) \sum_{j=1}^{\widetilde{T}_n} \frac{|\langle \phi, \psi_j\rangle|^2}{\lambda_j}.
\end{aligned}$$

This estimate, despite its technical simplicity, can be seen as the main idea of this section. It shows how to use regularization in the form of a truncation of the eigenvalues to obtain stable approximations for the sequence

$$n \mapsto \sum_{j=1}^{\widetilde{T}_n} \frac{|\langle \phi, \psi_j\rangle|^2}{\lambda_{j,n}}, \qquad n \in \mathbb{N}.$$

In the same way, one can estimate the series from below,

$$\begin{aligned}\sum_{j=1}^{\widetilde{T}_n} \frac{|\langle\phi,\psi_j\rangle|^2}{\lambda_{j,n}} &\geq \sum_{j=1}^{\widetilde{T}_n} \frac{|\langle\phi,\psi_j\rangle|^2}{\lambda_j + d_n} \\ &= \sum_{j=1}^{\widetilde{T}_n} \frac{|\langle\phi,\psi_j\rangle|^2}{\lambda_j} + \sum_{j=1}^{\widetilde{T}_n} \frac{|\langle\phi,\psi_j\rangle|^2 (\lambda_j - \lambda_j - d_n)}{\lambda_j(\lambda_j + d_n)} \\ &\geq \left(1 - \frac{1}{2+\gamma}\right) \sum_{j=1}^{\widetilde{T}_n} \frac{|\langle\phi,\psi_j\rangle|^2}{\lambda_j}\end{aligned}$$

since $\lambda_j + d_n \geq \lambda_{j,n} \geq (2+\gamma)d_n$. We already explained above that $\widetilde{T}_n \to \infty$ as $n \to \infty$. Hence, we obtain that the sequence $n \mapsto \sum_{j=1}^{\widetilde{T}_n} |\langle\phi,\psi_j\rangle|^2 / \lambda_{j,n}$ is bounded if and only if the sequence $n \mapsto \sum_{j=1}^{\widetilde{T}_n} |\langle\phi,\psi_j\rangle|^2 / \lambda_j$ is bounded, that is, if and only if $y \in D$. Our aim is now to perform a similar analysis when both the eigenvectors and eigenvalues enter the Picard series in a perturbed way. To deal with perturbations of the eigenvectors it will be helpful to consider the numbers

$$\rho_{k,n} := \min\{|\lambda_{k,n} - \lambda_{j,n}| : |\lambda_{k,n} - \lambda_{j,n}| > 2d_n,\ j \in \mathbb{N}\}, \qquad k, n \in \mathbb{N},$$

which indicate the distance of $\lambda_{k,n}$ to that part of the spectrum of $F_\sharp^n$ which does not converge to the same eigenvalue of $F_\sharp$ as $n \to \infty$: Since

$$|\lambda_{k,n} - \lambda_{j,n}| \leq |\lambda_{k,n} - \lambda_k| + |\lambda_k - \lambda_j| + |\lambda_j - \lambda_{j,n}| \leq 2d_n + |\lambda_k - \lambda_j|$$

for $j, k, n \in \mathbb{N}$, we see that if $|\lambda_{k,n} - \lambda_{j,n}| > 2d_n$, then $\lambda_{k,n}$ and $\lambda_{j,n}$ cannot be perturbations of the same eigenvalue. Stated in a different way, these two eigenvalues of $F_\sharp^n$ converge to two different eigenvalues of $F_\sharp$.

Using these prerequisites we define, for fixed parameters $\eta > 0$ and $0 < \vartheta < 1/2$ (which plays the role of an exponent in the following)

$$T_n = \max\{k \in \mathbb{N} : \lambda_{k,n} \geq (2+\eta)d_n^\vartheta,\ \rho_{k,n} \geq 4d_n^\vartheta,\ kd_n^{1-2\vartheta} < 1\}. \qquad \text{(I.43)}$$

Note that T_n does no longer depend on $F_\sharp$, but entirely on $F_\sharp^n$ and the noise level d_n. Moreover, as $d_n^\vartheta \to 0$, $\lambda_{k,n} \to \lambda_k$, and $\rho_{k,n} \to \operatorname{dist}(\lambda_k, \sigma(F_\sharp) \setminus \{\lambda_k\})$, we observe that $T_n \to \infty$ as $n \to \infty$. The following theorem is the main result of this section.

THEOREM I.12. *Let $\eta > 0$ and $\vartheta \in (0, 1/2)$ be arbitrary fixed parameters and denote by $(\lambda_{j,n}, \psi_{j,n})_{j\in\mathbb{N}}$ an eigensystem of the approximation $F_\sharp^n$ of $F_\sharp$ for $n \in \mathbb{N}$. Then the sequence*

$$n \mapsto \sum_{j=1}^{T_n} \frac{|\langle \phi, \psi_{j,n}\rangle|^2}{\lambda_{j,n}} \textit{ is bounded} \quad \Longleftrightarrow \quad \phi \in \mathrm{Range}(F_\sharp^{1/2}) \tag{I.44}$$

Proof. By definition of the truncation index T_n, it holds that $\lambda_j - d_n \geq \lambda_{j,n} - 2d_n \geq \eta d_n^\vartheta$ for $j \leq T_n$ and n so large that $d_n < 1$. Then

$$\begin{aligned}
\sum_{j=1}^{T_n} \frac{|\langle \phi, \psi_{j,n}\rangle|^2}{\lambda_{j,n}} &\leq \sum_{j=1}^{T_n} \frac{|\langle \phi, \psi_{j,n}\rangle|^2}{\lambda_j - d_n} \\
&= \sum_{j=1}^{T_n} \frac{|\langle \phi, \psi_j\rangle|^2}{\lambda_j} + d_n \sum_{j=1}^{T_n} \frac{|\langle \phi, \psi_j\rangle|^2}{\lambda_j(\lambda_j - d_n)} + \sum_{j=1}^{T_n} \frac{|\langle \phi, \psi_{j,n}\rangle|^2 - |\langle \phi, \psi_j\rangle|^2}{\lambda_j - d_n} \\
&\leq \left(1 + \frac{d_n^{1-\vartheta}}{\eta}\right) \sum_{j=1}^{T_n} \frac{|\langle \phi, \psi_j\rangle|^2}{\lambda_j} \\
&\quad + \left| \sum_{\lambda_j,\, j\leq T_n} \frac{1}{\lambda_j - d_n} \sum_{\lambda_l = \lambda_j} |\langle \phi, \psi_{l,n}\rangle|^2 - |\langle \phi, \psi_l\rangle|^2 \right|.
\end{aligned}$$

Here we exploited in the last step Lemma I.11, stating that the eigenspaces of $F_\sharp$ and $F_\sharp^n$ associated with λ_j and $\lambda_{j,n}$, respectively, have the same dimension. The last term in the last line of this estimate can be reformulated using the difference of the spectral projection of ϕ on the span of eigenvectors of $F_\sharp$ and $F_\sharp^n$ associated with λ_j and $\lambda_{j,n}$, respectively. We denote these projections by $P_j(F_\sharp)$ and $P_j(F_\sharp^n)$, respectively. Then we observe that

$$\begin{aligned}
&\sum_{\lambda_j,\, j\leq T_n} \frac{1}{\lambda_j - d_n} \sum_{\lambda_l = \lambda_j} |\langle \phi, \psi_l\rangle|^2 - |\langle \phi, \psi_{l,n}\rangle|^2 \\
&\qquad = \sum_{\lambda_j,\, j\leq T_n} \frac{1}{\lambda_j - d_n} \left(\|P_j(F_\sharp)\phi\|^2 - \|P_j(F_\sharp^n)\phi\|^2\right)
\end{aligned}$$

and hence

$$\left| \sum_{\lambda_j\,,\, j \le T_n} \frac{1}{\lambda_j - d_n} \sum_{\lambda_l = \lambda_j} |\langle \phi, \psi_l \rangle|^2 - |\langle \phi, \psi_{l,n} \rangle|^2 \right| \\ \le \frac{1}{\eta d_n^{\vartheta}} \sum_{\lambda_j\,,\, j \le T_n} \left| \|P_j(F_\sharp)\phi\|^2 - \|P_j(F_\sharp^n)\phi\|^2 \right| .$$

A simple computation shows that

$$\begin{aligned} &\|P_j(F_\sharp)\phi\|^2 - \|P_j(F_\sharp^n)\phi\|^2 \\ &\quad = \left(\|P_j(F_\sharp)\| - \|P_j(F_\sharp^n)\|\right)\left(\|P_j(F_\sharp)\| + \|P_j(F_\sharp^n)\|\right)\|\phi\|^2 \\ &\quad \le \left| \|P_j(F_\sharp)\| - \|P_j(F_\sharp^n)\| \right| 2\|\phi\|^2 \le \|P_j(F_\sharp) - P_j(F_\sharp^n)\|\, 2\|\phi\|^2 \end{aligned}$$

since the operator norms of the orthogonal projectors $P_j(F_\sharp)$ and $P_j(F_\sharp^n)$ are bounded by one. Next we estimate the norm $\|P_j(F_\sharp) - P_j(F_\sharp^n)\|$. As we have explained above, the spectral projections can be written as

$$P_j(A) = \frac{1}{2\pi \mathrm{i}} \int_{\gamma_j} R(\xi, A)\, \mathrm{d}\xi \qquad \text{where } A = F_\sharp \text{ or } A = F_\sharp^n. \tag{I.45}$$

We still have to define the curve γ_j, merely using $F_\sharp^n$, and the problem is here that multiple eigenvalues of $F_\sharp$ might result in different eigenvalues of $F_\sharp^n$. Therefore we proceed as follows: To each $\lambda_{j,n} \in \sigma(F_\sharp^n)$ we associate its cluster

$$\mathrm{clu}\,(\lambda_{j,n}) := \{\lambda_{k,n} : |\lambda_{k,n} - \lambda_{j,n}| \le 2d_n\}, \qquad j \in \mathbb{N},\, n \in \mathbb{N}.$$

It is clear that $\lambda_{j,n} \in \mathrm{clu}\,(\lambda_{k,n})$ precisely if $\lambda_{k,n} \in \mathrm{clu}\,(\lambda_{j,n})$. Moreover, if $j, k, l \le T_n$ and n is large enough such that $d_n < 1$, then the cluster determines even a transitive relation: If $\lambda_{j,n} \in \mathrm{clu}\,(\lambda_{k,n})$ and $\lambda_{k,n} \in \mathrm{clu}\,(\lambda_{l,n})$ then $\lambda_{j,n} \in \mathrm{clu}\,(\lambda_{l,n})$. Indeed, otherwise, by the definition of T_n,

$$4d_n^{\vartheta} \le \rho_{j,n} \le |\lambda_{j,n} - \lambda_{l,n}| \le |\lambda_{j,n} - \lambda_{k,n}| + |\lambda_{k,n} - \lambda_{l,n}| \le 4d_n,$$

which is a contradiction for n so large that $d_n < 1$. Hence, for fixed n we can partition the first T_n eigenvalues in groups of clusters: There is $m : \mathbb{N} \to \mathbb{N}$ and $I_n \le T_n$ such that $\bigcup_{j \le I_n} \mathrm{clu}\left(\lambda_{m(j),n}\right) = \{\lambda_{j,n} : j = 1, \ldots T_n\}$

and the union is disjoint. Around each $\lambda_{m(j),n}$ we consider the circle $\Gamma_{j,n}$ with radius $2d_n^\vartheta$. Then all elements in $\mathrm{clu}\left(\lambda_{m(j),n}\right)$ lie inside $\Gamma_{j,n}$ and at least one eigenvalue of $F_\sharp$, too, since $\mathrm{dist}(\sigma(F_\sharp), \sigma(F_\sharp^n)) \leq d_n$. Moreover, the circles $\Gamma_{j,n}$ do not intersect, since otherwise there exist $\lambda_{k,n}$ and $\lambda_{l,n}$, $1 \leq k, l \leq T_n$ with $2d_n < |\lambda_{k,n} - \lambda_{l,n}| \leq 4d_n^\vartheta$, which is impossible by definition of T_n, see (I.43). Finally, we note that

$$\mathrm{dist}\left(\Gamma_{j,n}, \sigma(F_\sharp^n)\right) \geq 2d_n^\vartheta - 2d_n \geq Cd_n^\vartheta \quad \text{and}$$
$$\mathrm{dist}\left(\Gamma_{j,n}, \sigma(F_\sharp)\right) \geq 2d_n^\vartheta - 3d_n \geq Cd_n^\vartheta$$

for n large enough. Now, the positively oriented curve γ_n is defined such that its path is the union of the circles $\Gamma_{j,n}$ of radius $2d_n^\vartheta$. If follows from the last equation that the integral representing $P_j(F_\sharp^n)$ in (I.45) is now well defined. Moreover, due to Lemma I.11 we can estimate

$$\|P_j(F_\sharp) - P_j(F_\sharp^n)\| \leq \frac{d_n}{2d_n^\vartheta - 3d_n - d_n} \leq Cd_n^{1-\vartheta}$$

for some constant C and n large enough. We conclude that

$$\left| \sum_{\lambda_j,\, j \leq T_n} \frac{1}{\lambda_j - d_n} \sum_{\lambda_l = \lambda_j} |\langle\phi, \psi_l\rangle|^2 - |\langle\phi, \psi_{l,n}\rangle|^2 \right| \leq CT_n d_n^{1-\vartheta}$$

and, collecting terms,

$$\begin{aligned}
\sum_{j=1}^{T_n} \frac{|\langle\phi, \psi_{j,n}\rangle|^2}{\lambda_{j,n}} &\leq \left(1 + \frac{d_n^{1-\vartheta}}{\eta}\right) \sum_{j=1}^{T_n} \frac{|\langle\phi, \psi_j\rangle|^2}{\lambda_j} + CT_n \frac{d_n^{1-2\vartheta}}{\eta} \|\phi\|^2 \\
&\leq \left(1 + \frac{d_n^{1-\vartheta}}{\eta}\right) \sum_{j=1}^{T_n} \frac{|\langle\phi, \psi_j\rangle|^2}{\lambda_j} + \frac{C}{\eta} \|\phi\|^2
\end{aligned}$$

where we exploited again the definition of T_n in (I.43) in the last step. Obviously, the left hand side is uniformly bounded in n if the sequence $n \mapsto \sum_{j=1}^{T_n} |\langle\phi, \psi_j\rangle|^2 / \lambda_j$ is bounded.

An estimate of the perturbed Picard series from below is achieved analogously. Since $\lambda_{j,n} \leq \lambda_j + d_n$ and $\lambda_{j,n} + d_n \leq \lambda_{j,n} + 2d_n \leq (2+\eta)d_n^\vartheta + 2d_n \leq$

$(4+\eta)d_n^\vartheta$ it holds that

$$\sum_{j=1}^{T_n} \frac{|\langle \phi, \psi_{j,n}\rangle|^2}{\lambda_{j,n}} \geq \sum_{j=1}^{T_n} \frac{|\langle \phi, \psi_{j,n}\rangle|^2}{\lambda_j + d_n}$$

$$= \sum_{j=1}^{T_n} \frac{|\langle \phi, \psi_j\rangle|^2}{\lambda_j} - d_n \sum_{j=1}^{T_n} \frac{|\langle \phi, \psi_j\rangle|^2}{\lambda_j(\lambda_j + d_n)} + \sum_{j=1}^{T_n} \frac{|\langle \phi, \psi_{j,n}\rangle|^2 - |\langle \phi, \psi_j\rangle|^2}{\lambda_j + d_n}$$

$$\geq \left(1 + \frac{d_n^{1-\vartheta}}{4+\eta}\right) \sum_{j=1}^{T_n} \frac{|\langle \phi, \psi_j\rangle|^2}{\lambda_j} - CT_n \frac{d_n^{1-2\vartheta}}{4+\eta} \|\phi\|^2$$

$$\geq \left(1 + \frac{d_n^{1-\vartheta}}{4+\eta}\right) \sum_{j=1}^{T_n} \frac{|\langle \phi, \psi_j\rangle|^2}{\lambda_j} - \frac{C}{4+\eta} \|\phi\|^2.$$

Consequently, if the sequence $n \mapsto \sum_{j=1}^{T_n} |\langle \phi, \psi_j\rangle|^2/\lambda_j$ diverges for some $y \in \mathbb{R}^2$, then $n \mapsto \sum_{j=1}^{T_n} |\langle \phi, \psi_{j,n}\rangle|^2/\lambda_{j,n}$ diverges. Since $T_n \to \infty$ as $n \to \infty$ we obtain the claim of the theorem. □

In view of the Factorization method, the preceding theorem shows that it is reasonable to believe that a truncated series criterion delivers meaningful reconstructions for perturbed data, when the truncation index is chosen in a suitable regularizing way. This explains why the Factorization method works in practice, when only perturbed finite dimensional data are at hand. In the following corollary, we strengthen the convergence properties of the perturbed regularized Picard series. Therefore we use the reciprocal of the Picard series,

$$y \mapsto \left(\sum_{j=1}^{T_n} \frac{|\langle \Phi_\infty(\cdot, y), \psi_{j,n}\rangle|^2}{\lambda_{j,n}} \right)^{-1}.$$

which is often plotted for some fixed n as a function of y in numerical validations of the Factorization method.

Corollary I.13. *Let $K \subset \mathbb{R}^2 \setminus D$ be compact. Then*

$$\sup_{y \in K} \left(\sum_{j=1}^{T_n} \frac{|\langle \Phi_\infty(\cdot, y), \psi_{j,n}\rangle|^2}{\lambda_{j,n}} \right)^{-1} \to 0 \qquad \text{as } n \to \infty.$$

Proof. Since the norms $\|\Phi_\infty(\cdot, y)\|_{L^2(\mathbb{S}1)}$ are uniformly bounded in $y \in \mathbb{R}^2$, we know from the proof of Theorem I.12 that there are constants C_1, C_2 such that

$$0 \leq \left(\sum_{j=1}^{T_n} \frac{|\langle \phi_y, \psi_{j,n} \rangle|^2}{\lambda_{j,n}} \right)^{-1} \leq C_1 \left(\sum_{j=1}^{T_n} \frac{|\langle \phi_y, \psi_j \rangle|^2}{\lambda_j} - C_2 \right)^{-1} =: S_n(y). \tag{I.46}$$

By construction of T_n, the series on the right hand side tends to 0 as $n \to \infty$ pointwise for every $y \in K$. Moreover, the right hand side is monotonous in n, $0 < S_{n+1}(y) \leq S_n(y)$ for $y \in K$ and $n \in \mathbb{N}$ and each partial sum of the Picard series is continuous in y as a finite sum of continuous functions. Hence, Dini's theorem [84] states that $S_n(y) \to 0$ uniformly in K, which yields the claim of the corollary due to (I.46). □

We finish this chapter with several numerical examples of domain reconstructions by the Factorization method. The data we use is generated by an implementation of the Lippmann-Schwinger equation (I.19) based on fast Fourier transform techniques which is described in [78] and goes back to [83], see also [77]. The first inclusion we consider is kite-shaped, the value of the refractive index inside the kite is equal to ten and outside equal to one. The wavenumber k in all our examples below is equal to one. We approximate the far field operator using 32 incident angles and measurement directions equally distributed on the unit circle according to (I.33). A glance at the magnitude of the eigenvalues in Figure I.1 of the approximation corresponding to the auxiliary operator $F_\sharp$ shows exponential decay of the eigenvalues round about up to the 8th eigenvalue and then a tail of eigenvalues which are of the order $5 \cdot 10^{-14}$. Except for special cases, the precise noise level d_{32} from (I.42) is of course unknown, but the precision of the forward data can be estimated by the order of magnitude of the eigenvalues in the tail of the eigenvalue sequence of the approximation to $F_\sharp$. Hence, as a (brute) approximation of the cut-off criterion (I.43) we only take those eigenvectors and values in consideration for which the magnitude of the eigenvalue is larger than the noise level of the forward data. The precise cut-off parameter is determined "by hand", with a glance at the spectrum of the corresponding matrix. This is of course no guarantee that the corresponding eigenvectors have a similar precision, but it is unclear to us how to set up an analogous criterion for the eigenspaces. Several authors investigated the choice of the cut-off index before, see for instance [33] for a related technique, but also [40, 58]. In Figure I.1 we plot the magnitude of the spectrum of our numerical approximation to $F_\sharp$ where no artificial noise was added. Round about nine eigenvalues seem to be above the noise level. Figure I.2 shows the corresponding reconstruction of the support of the scatterer plotting the reciprocal of the function in (I.44) for $T_{32} = 9$ and $n = 32$. The reconstruction looks roughly like a triangle, the non-convexity of the scattering object is hardly observable. Note, however, that the wavelength $2\pi/k$ in the background equals 2π and is longer than the diagonal of the plot, which has length and height equal to four. It is hence not surprising that small details of the scattering object are not resolved. In [66, Section 5.4] it is remarked that the reconstruction of an inhomogeneous medium using the reciprocity gap functional is less satisfactory than for a perfect conductor, and the same has been observed for the Linear Sampling method in [20]. Comparing our results to examples of reconstructions of impenetrable scatterers by the Factorization method given in [49], we note in analogy to the just mentioned references that our medium reconstructions have less quality.

Figure I.3 shows again the spectrum of the numerical approximation to $F_\sharp$, where the far field pattern is now perturbed by one percent artificial (relative) noise, measured in the Frobenius norm. Round about 4 modes still seem to be precise, whereas the remaining tail of the eigenvalues are all completely perturbed. Figure I.4 shows the corresponding reconstruction using the reciprocal of the criterion (I.44) for $T_{32} = 4$. Now merely the position of the scatterer is observable and the shape is not reconstructed anymore. However, note that only four modes have been used for this reconstruction.

Now we consider a disconnected scatterer consisting of one kite and one smaller ball above the kite. The value of the refractive index inside the inhomogeneity equals as for the single kite ten in both parts of the obstacle. Again, we plot the magnitude of the eigenvalues of the approximation to $F_\sharp$ without artificial noise, see Figure I.5, observe that we can trust the eight first modes and use these for a reconstruction of the scatterer's support, shown in Figure I.6. The reconstruction shows the two components of the scatterer and the shape of the kite is well approximated. There appears a kind of bridge between the scatterers, which is not surprising since the wavelength 2π is again larger than the diagonal

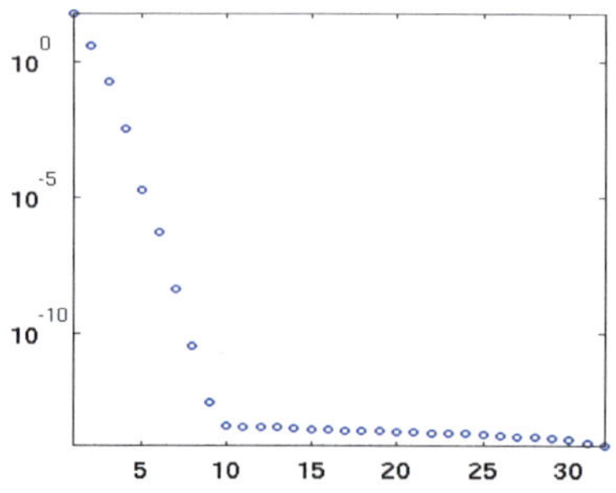

Figure I.1: Magnitude of eigenvalues of the approximation to $F_\sharp$.

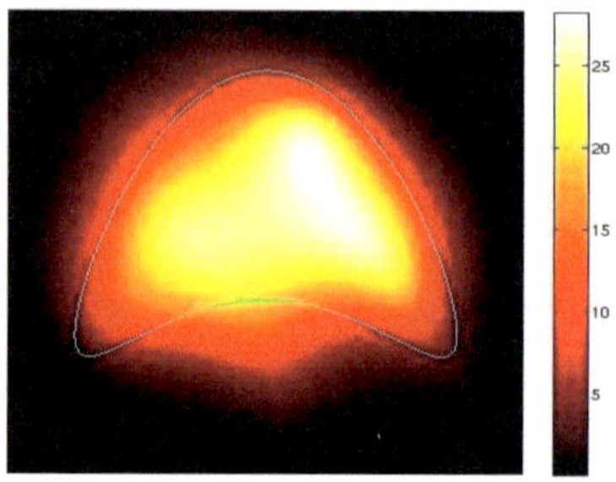

Figure I.2: $T_{32} = 9$, no artificial noise.

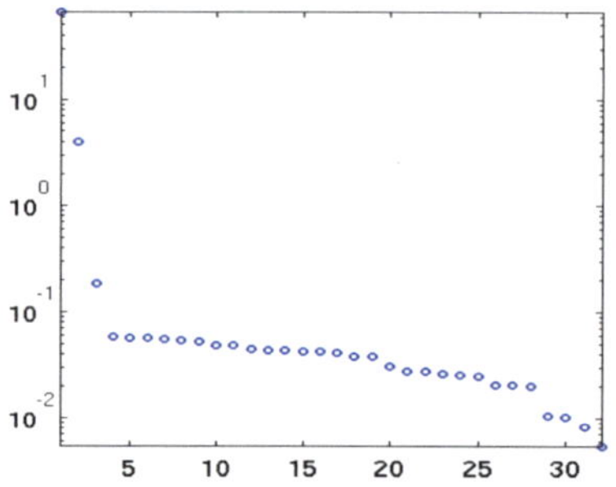

Figure I.3: Magnitude of eigenvalues of the approximation to $F_\sharp$, one percent artificial noise.

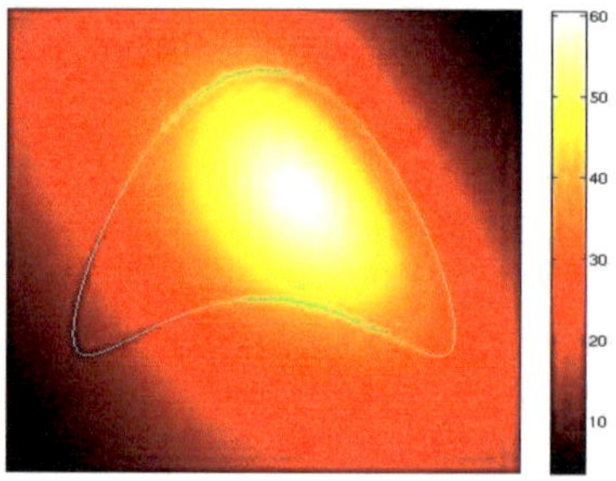

Figure I.4: $T_{32} = 4$, one percent artificial noise.

Figures I.1–I.4: Reconstructions of the support of a piecewise constant refractive index.

of the plot of length $4\sqrt{2}$. We perturb the approximation to the far field operator by one percent relative artificial noise in the Frobenius norm and note that merely three modes seem to be acceptable, compare Figure I.7. The corresponding reconstruction is shown in Figure I.8, and shows again good localization of the scattering object, but no separation of its two components.

In implementations of the Factorization method one observes sometimes that eigenpairs of the approximation to $F_\sharp$ for which the noise level of the data is of the order of magnitude of the eigenvalue still improve the reconstruction considerably. Such eigenvalues are of course useless from the numerical point of view, as they represent by no means an approximation to the true eigenvalue, and the same holds for the corresponding eigenspace. The above theory tells nothing about this phenomenon, however, we believe that the concept of the so-called *noise subspace*, presented in [62], might give some explanation of this phenomenon, which is also present here: Plotting the reciprocal of the sum in (I.44) for the same data as in Figure I.7 for $T_{32} = 29$ shows two separated parts of the scatterer in Figure I.9. The value 29 has been chosen rather arbitrarily, the same procedure for $T_{32} = 28$ yields in Figure I.10 an even sharper separation of the two objects.

Finally, we consider a partially absorbing scattering object with two components which are the same kite and ball as in the previous experiment, but the contrast of the material is now vanishing at one of the scatterer's boundaries. This time, the refractive index inside the ball equals ten whereas inside the kite the refractive index is given in polar coordinates (r, ϕ) by $1+4(1+i)x(\phi)\big((r\cos(\phi)-0.8)^2+(r\sin(\phi)+0.6)^2-0.5\big)^2$, where $x(\phi)$, $\phi \in [0, 2\pi]$, is a parameterization of the kite's boundary, analogous to the one in [22, Page 70]. In particular, the corresponding contrast to this refractive index vanishes at the boundary of the kite. As above, after a look at the magnitude of the spectral values of the approximation to $F_\sharp$, see Figure I.11, we decide to trust the first eight modes and plot the corresponding reconstruction in Figure I.12. The quality of the reconstruction of the support is comparable to that of the piecewise constant medium in Figure I.6. Again, we perturb the approximation to the auxiliary operator $F_\sharp$ by one percent relative noise in the Frobenius norm. Using merely the three modes above the noise level the Factorization method is again only able to find the location of the scatterer, but not the number of connected components. As in the previous example, we see again that taking into account more modes (which represent by no means an approximation to the true eigenvalues) produces reconstructions which show the two components of the scatterer, however, as above not their shape.

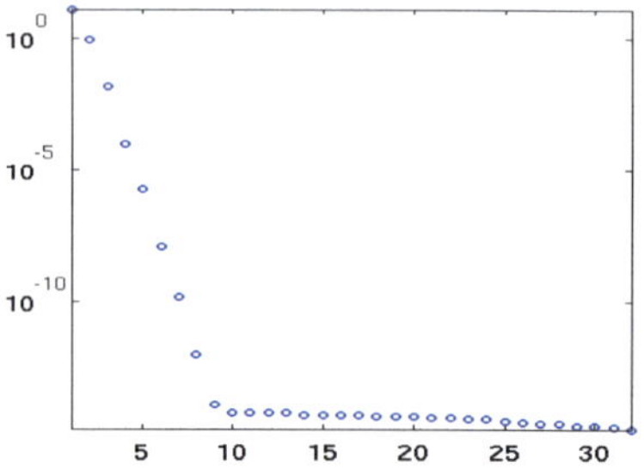

Figure I.5: Magnitude of eigenvalues of the approximation to $F_\sharp$

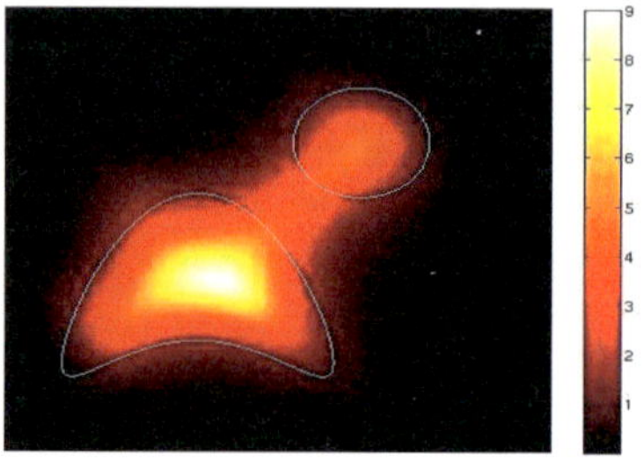

Figure I.6: $T_{32} = 8$, no artificial noise.

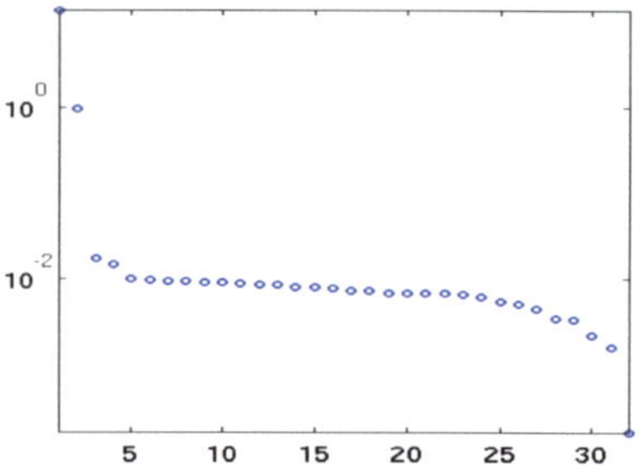

Figure I.7: Magnitude of eigenvalues of the approximation to $F_\sharp$, one percent artificial noise.

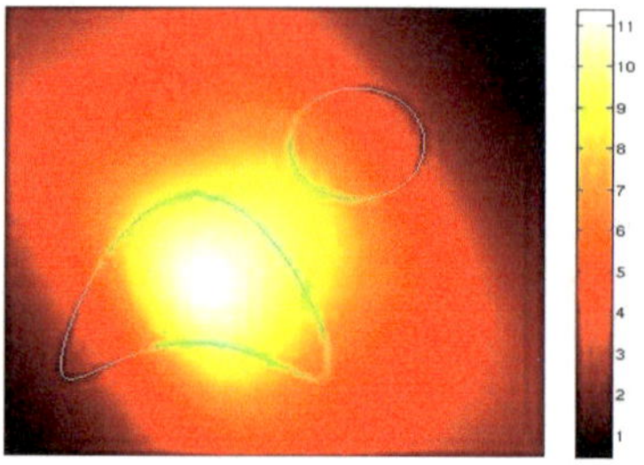

Figure I.8: $T_{32} = 4$, one percent artificial noise.

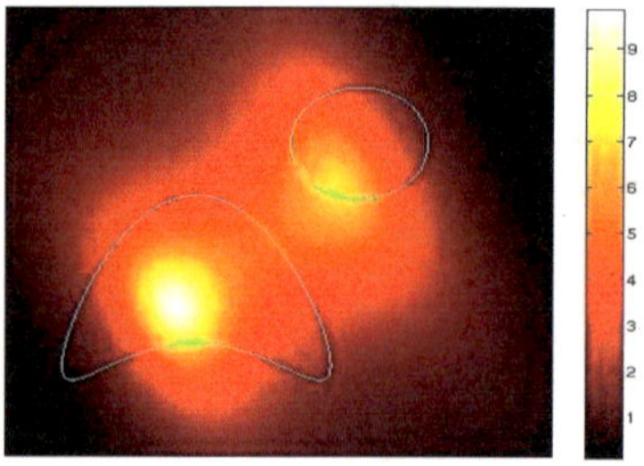

Figure I.9: $T_{32} = 29$, one percent artificial noise.

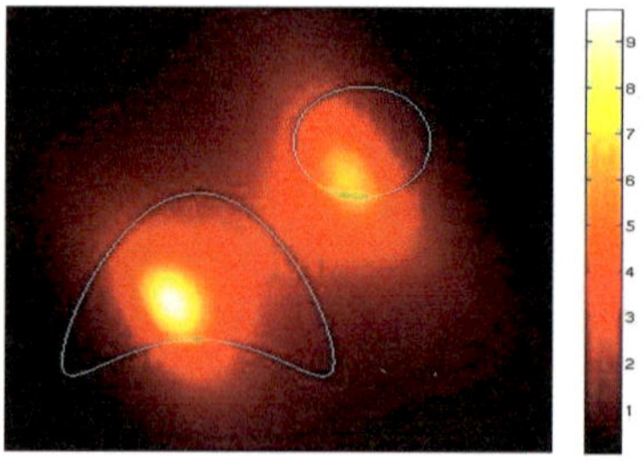

Figure I.10: $T_{32} = 28$, one percent artificial noise.

Figures I.5–I.10: Reconstructions of the disconnected support of a piecewise constant refractive index.

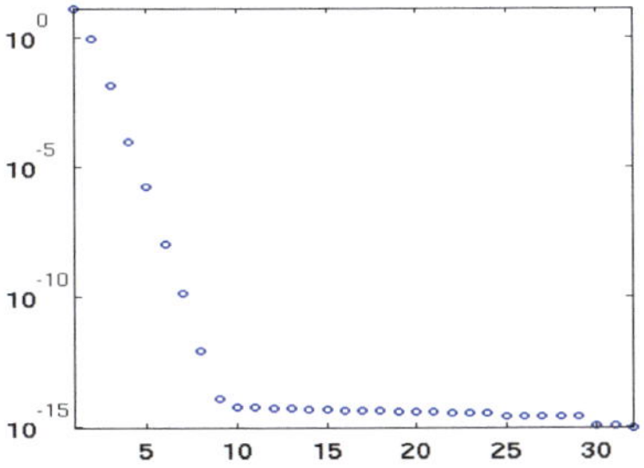

Figure I.11: Magnitude of eigenvalues of the approximation to $F_\sharp$

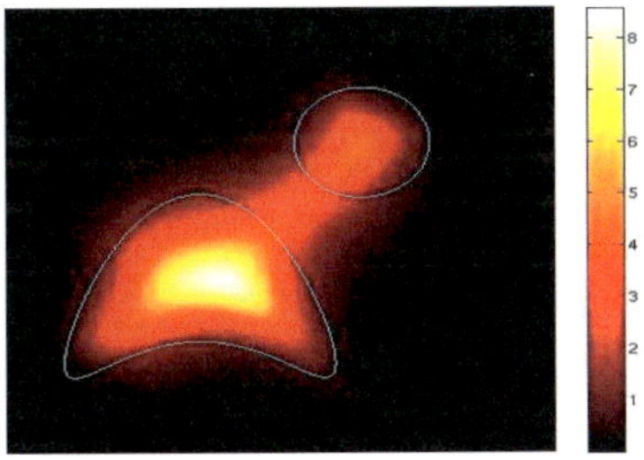

Figure I.12: $T_{32} = 8$, no artificial noise.

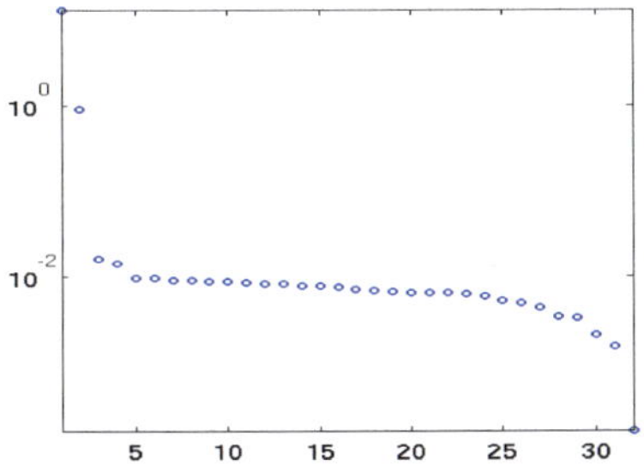

Figure I.13: Magnitude of eigenvalues of the approximation to $F_\sharp$, one percent artificial noise.

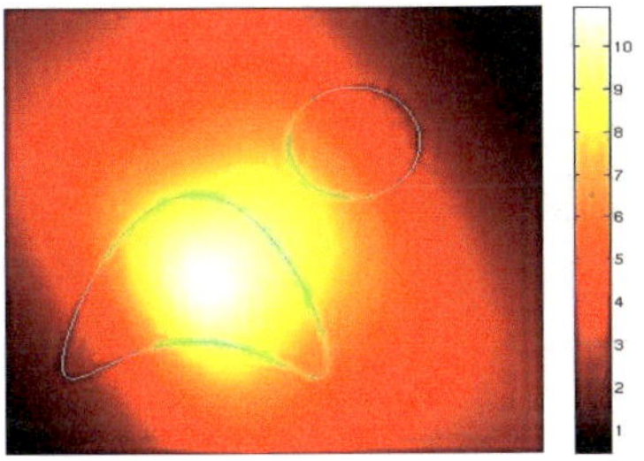

Figure I.14: $T_{32} = 3$, one percent artificial noise.

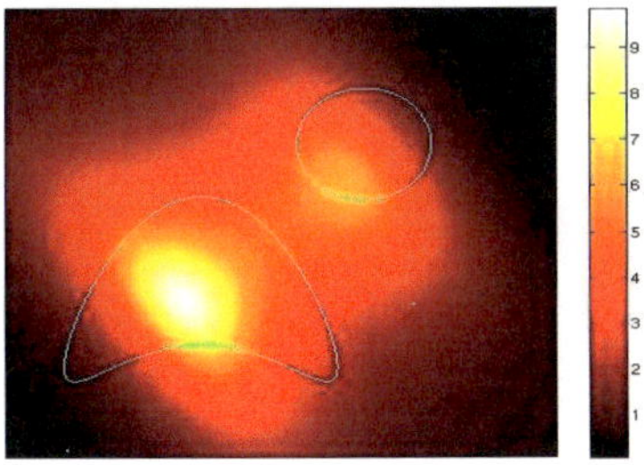

Figure I.15: $T_{32} = 29$, one percent artificial noise.

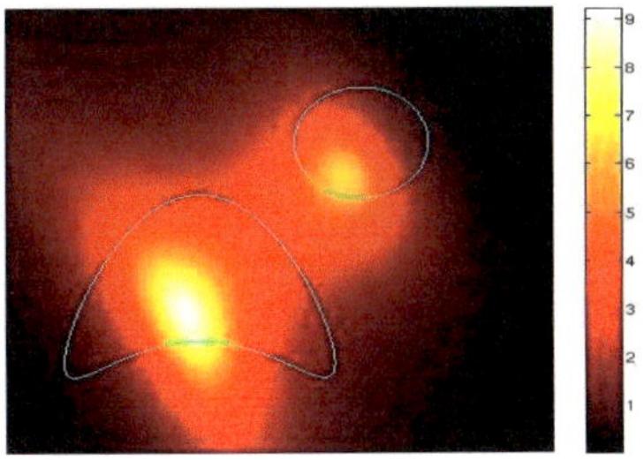

Figure I.16: $T_{32} = 27$, one percent artificial noise.

Figures I.11–I.16: Reconstructions of the support of a smooth and partly complex valued refractive index.

CHAPTER II

The Factorization Method for Photonic Crystals

II-1. Introduction

In this chapter, we develop a Factorization method for the inverse transmission problem of reconstructing a periodic interface between two media from measurements of scattered fields in a finite distance above the interface. The periodic and binary two dimensional material configuration, also called a diffraction grating, is a simple model for a photonic crystal – a dielectric structure with periodically varying material parameters taking only two values. The direct transmission problem arises for instance when monochromatic light is scattered at a photonic crystal or at a diffraction grating made of dielectrics. Possible applications of such materials include splitting and guiding of light in nano scale devices. The inverse problem arises naturally as an identification or non-destructive testing problem in nano optics.

For the inverse periodic transmission problem (which is stated in a more precise form on page 64) we apply the Factorization method in a similar way as we applied it in Chapter I to the inverse medium problem, and derive an explicit characterization of the periodic interface. However, the inverse transmission problem for periodic structures is more difficult to treat, since "standard" assumptions for the middle operator of the corresponding near field factorization $N = -LM^*L^*$ fail: M is without suitable modification indefinite. However, following [17], a suitably modified version of the selfadjoint part of M is a compact perturbation of a coercive operator. We will observe later on that the imaginary part of the middle operator M^* is merely semidefinite and hence our generalized version of the basic result on range identities of the Factorization method in Theorem I.7 makes the construction of the method convenient. To emphasize this fact, we also show the tremendous effort necessary for the construction of the method when only an earlier version due to Kirsch and Grinberg, see [49, 47], is at hand. In that case, we have to combine holomorphic Fredholm theory with the Factorization method, and to pay the price of possibly loosing exceptional frequencies where the method cannot be proven to work. By holomorphic Fredholm theory we conclude that $\mathrm{Re}\,\mathcal{M}$ is injective for all but a countable sequence of real frequencies and construct a Factorization method using the "old" result on range identities. Using our new version, we do not loose exceptional frequencies during the treatment of the inverse problem and avoid the application of holomorphic Fredholm theory. We finally show how to close a gap in former applications [7, 3] of the method to periodic structures where Dirichlet boundary conditions have been treated.

In the remainder of this introductory section, we present the setting for the direct scattering problem of this chapter in detail. We consider time harmonic scattering from a periodic interface Γ in $\mathbb{R}^3$ that represents the interface between two media with constant, but different, refractive indices. The interface Γ is assumed to be sufficiently smooth (more precisely, of class $C^{3,\alpha}$, $\alpha \in (0,1)$) and 2π periodic in x_1 direction, i.e., for $x \in \Gamma$ the point $x + j(2\pi, 0, 0)$ also belongs to Γ for $j \in \mathbb{Z}$. In x_3 direction, the interface is constant. We fix some number $h > 0$ such that $0 < x_2 < h$ for $x \in \Gamma$.

As we derived in Section I-2, time harmonic scattering of electromagnetic plane waves at Γ reduces for the transverse electric and magnetic mode to a scalar partial differential equation posed on a two dimensional domain. We only consider such a reduced problem and remark that, despite our restriction to the two dimensional case, all results can be extended to scattering from biperiodic media in three dimensions. Let us from now on identify the interface Γ with its restriction to the two dimensional "unit cell"

$$\Pi = (0, 2\pi) \times \mathbb{R} \subset \mathbb{R}^2$$

and work in a two dimensional setting, compare Figure II.1. We introduce the two segments

$$\Gamma_+ := \{x \in \Pi : x_2 = h\} \qquad \text{and} \qquad \Gamma_- := \{x \in \Pi : x_2 = 0\}$$

and denote by $\Omega_+ \subset \Pi$ the set of all points x "above" Γ and by $\Omega_- \subset \Pi$ the domain "below" Γ. More precisely, a point belongs to Ω_+ if it is connected to Γ_+ by some continuous path that does not cross the interface Γ and $\Omega_- := \Pi \setminus \overline{\Omega_+}$. We require that Ω_+ is connected, but Ω_- may consist of several components. We define $D_\pm$ to be two domains of finite height above and below Γ,

$$D_+ := \{x \in \Omega_+,\, x_2 < h\}, \qquad D_- := \{x \in \Omega_-,\, x_2 > 0\},$$

and set

$$D = D_+ \cup \Gamma \cup D_- = (0, 2\pi) \times (0, h).$$

On the curve Γ we define a unit normal field ν that points into Ω_+ and on $\Gamma_\pm$ we choose ν to be exterior to $D_\pm$. Note that we do *not* require Γ to be the graph of a function.

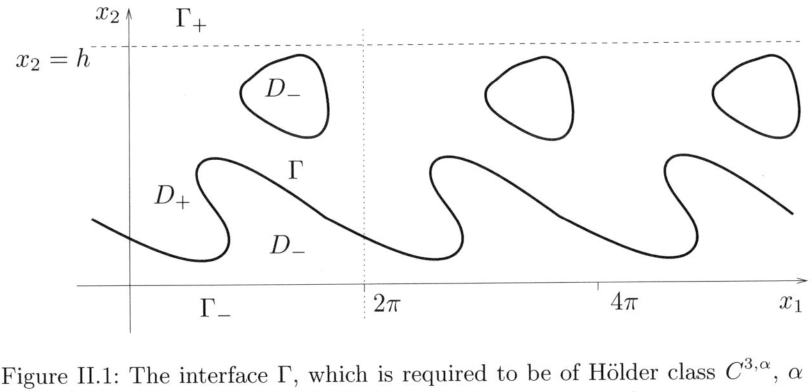

Figure II.1: The interface Γ, which is required to be of Hölder class $C^{3,\alpha}$, $\alpha \in (0,1)$. The upper domain D_+ is supposed to be connected.

Physical properties of the dielectric $\Omega_\pm$ are modelled by the permittivity ε which takes constant values $\varepsilon_+ > 0$ in Ω_+ and $\varepsilon_- > 0$ in Ω_-. These two constants are fixed in the entire chapter and determine the piecewise constant function

$$\varepsilon := \begin{cases} \varepsilon_+ & \text{in } \Omega_+, \\ \varepsilon_- & \text{in } \Omega_-. \end{cases}$$

The permeability μ is supposed to be a positive constant whereas the conductivity σ is assumed to vanish in this chapter. As in Chapter I, the number $\omega > 0$ denotes the angular frequency and the wavenumber $k_\pm$ is defined as $k_\pm = \omega\sqrt{\varepsilon_\pm \mu_0}$. Since we rely several times in this chapter explicitely on the frequency ω instead of the wavenumber k, it eases notation if we assume, without loss of generality, that $\mu_0 = 1$, thus,

$$k_\pm = \omega\sqrt{\varepsilon_\pm} \quad \text{and} \quad k := \begin{cases} k_+ & \text{in } \Omega_+, \\ k_- & \text{in } \Omega_-. \end{cases}$$

II-2. The Direct Transmission Problem

In general, incident plane waves do not share the 2π periodicity in x_1 direction of the scatterer Γ, but they are *α-quasiperiodic.* A function $f : \mathbb{R} \to \mathbb{C}$ is called quasiperiodic with phase shift $\alpha \in \mathbb{R}$ if $f(x + 2\pi) = \exp(2\pi \mathrm{i}\alpha) f(x)$, $x \in \mathbb{R}$. A function with domain of definition in $\mathbb{R}^2$ is called α-quasiperiodic if it is α-quasiperiodic in the first variable x_1. Note the following shorthand notation for the jump of a function across Γ: For $f : \Pi \to \mathbb{C}$ such that the traces $f|_\Gamma^+$ from Ω_+ and $f|_\Gamma^-$ from Ω_- on Γ are well defined, we set

$$[f]_\Gamma := f|_\Gamma^+ - f|_\Gamma^- .$$

The direct transmission scattering problem for the transverse magnetic mode is formulated as follows (compare Section I-1). For an incident plane wave propagating towards Γ from above,

$$u^i = e^{\mathrm{i}k_+ \hat{\theta}\cdot x}, \qquad \hat{\theta} = (\cos\theta, -\sin\theta), \qquad \theta \in (0, \pi),$$

which is quasiperiodic with phase shift $\alpha = k_+ \cos\theta$, we want to find an α-quasiperiodic total field u^t solving

$$\Delta u^t + \omega^2 \varepsilon u^t = 0 \qquad \text{in } \Omega_+ \cup \Omega_-. \tag{II.1}$$

At the interface Γ a pair of transmission conditions is imposed, compare [23],

$$\left[u^t\right]_\Gamma = 0 \quad \text{and} \quad \left[\frac{\partial u^t}{\partial \nu}\right]_\Gamma = 0. \tag{II.2}$$

This formulation of the direct problem does not yet guarantee uniqueness of solution. A radiation condition has to be imposed on the scattered field $u^s = u^t - u^i$ in Ω_+ and on u^t in Ω_-. In physical terms, such a condition guarantees that energy is transported away from the interface Γ. It is well known [23, 44] that the correct way of imposing a radiation condition for the problem at hand is to assume that the scattered field away from the crystal can be represented as a Rayleigh series

$$u^s(x) = \sum_{n\in\mathbb{Z}} u_n^+ e^{\mathrm{i}(\alpha_n x_1 + \beta_n^+ x_2)} \quad \text{for } x_2 > h,$$
$$\text{with } \alpha_n = n + \alpha, \quad \beta_n^+ = \sqrt{k_+^2 - \alpha_n^2} = \sqrt{\omega^2\varepsilon_+ - \alpha_n^2}, \quad \text{(II.3)}$$

which is required to converge uniformly on compact subsets of $\{x \in \mathbb{R}^2 : x_2 > h\}$. The square root in definition (II.3) is defined as the unique holomorphic extension of the square root on the positive numbers to all of $\mathbb{C} \setminus [0, -\mathrm{i}\infty)$, which is the complex plane slit at the negative imaginary axis. Especially, if ω is positive,

$$\beta_n^+ = \beta_n^+(\omega) = \begin{cases} \sqrt{\omega^2\varepsilon_+ - \alpha_n^2}, & |\alpha_n| \le \omega(\varepsilon_+)^{1/2}, \\ \mathrm{i}\sqrt{\alpha_n^2 - \omega^2\varepsilon_-}, & |\alpha_n| > \omega(\varepsilon_-)^{1/2}, \end{cases} \quad \text{for } n \in \mathbb{Z}. \quad \text{(II.4)}$$

Our definition of the square root necessitates to exclude certain frequencies ω, namely those for which $\omega^2\varepsilon_\pm - \alpha_n^2 \in [0, -\mathrm{i}\infty)$ for some $n \in \mathbb{Z}$. We denote this exceptional set by

$$\mathcal{E} = \{\omega \in \mathbb{C} : \omega^2\varepsilon_+ - \alpha_n^2 \text{ or } \omega^2\varepsilon_- - \alpha_n^2 \text{ belong to } [0, -\mathrm{i}\infty) \text{ for some } n \in \mathbb{Z}\}. \quad \text{(II.5)}$$

Note that $\mathcal{E} \cap \mathbb{R}_{>0} = \{\omega > 0 : \omega = |\alpha_n|\varepsilon_+^{-1/2} \text{ or } \omega = |\alpha_n|\varepsilon_-^{-1/2},\ n \in \mathbb{Z}\}$ consists of a sequence of numbers tending to infinity. We impose a similar condition as (II.3) on the total field u^t below Γ. More precisely,

$$u^t(x) = \sum_{n\in\mathbb{Z}} u_n^- e^{\mathrm{i}(\alpha_n x_1 - \beta_n^- x_2)}, \quad \text{for } x_2 < 0,$$
$$\text{where } \beta_n^- = \sqrt{k_-^2 - \alpha_n^2} = \sqrt{\omega^2\varepsilon_- - \alpha_n^2}. \quad \text{(II.6)}$$

Conditions (II.3), (II.6) are in the sequel referred to as the upper and lower Rayleigh conditions, respectively.

As in Chapter I we deal with weak solutions when solving the direct scattering problem and therefore need to introduce convenient function spaces. We set, for $\alpha \in \mathbb{R}$,

$$C_\alpha^\infty(D) := \big\{u \in C^\infty(D) : e^{-\mathrm{i}\alpha x_1} u(x) = U|_D \qquad \text{for an } x_1\text{-periodic } U \in C^\infty(\mathbb{R}^2)\big\}$$

and define $H_\alpha^1(D)$ to be the closure of $C_\alpha^\infty(D)$ in the $H^1(D)$ norm. The same construction works for $D_\pm$ or for Γ and one can also define spaces with local integrability as, e.g., $H^1_{\alpha,\mathrm{loc}}(\Pi)$ or $H^1_{\alpha,\mathrm{loc}}(\Omega_\pm)$. The trace space of $H_\alpha^1(D)$ is the periodic Sobolev space $H_\alpha^{1/2}(\Gamma)$, defined by $H_\alpha^s(\Gamma) := \overline{C_\alpha^\infty(\Gamma)}^{H^s(\Gamma)}$, $s \in \mathbb{R}$. We recall that $u \in H_\alpha^1(D_\pm)$ possesses a unique normal derivative $\partial u/\partial\nu \in H_\alpha^{-1/2}(\Gamma)$ if $\Delta u \in L^2(D_\pm)$ in the weak sense [63]. A variational formulation of the direct problem, originally posed on an unbounded domain, needs to incorporate the Rayleigh conditions (II.3) and (II.6) in some way. One option is to use transparent boundary conditions on the artificial boundaries $\Gamma_\pm$. For $\phi \in H_\alpha^{1/2}(\Gamma_+)$ the α-quasiperiodic solution v of the Helmholtz equation in Ω_+ that satisfies the Rayleigh condition (II.3) and takes boundary values $v = \phi$ on Γ_+ is

$$v(x) = \sum_{n\in\mathbb{Z}} \hat{\phi}_n e^{\mathrm{i}(\alpha_n x_1 + \beta_n^+(x_2-h))}, \qquad x_2 > h,$$

where $\hat{\phi}_n$ denotes the nth Fourier coefficient of the periodic function $\exp(-\mathrm{i}\alpha\,\cdot)\phi$. The normal derivative of v on Γ_+ is, at least formally, $\partial v/\partial\nu = \mathrm{i}\sum \beta_n^+ \hat{\phi}_n e^{\mathrm{i}\alpha_n x_1}$. This motivates to define the Dirichlet-to-Neumann (or Steklov–Poincaré) operator

$$\Lambda^+ : \phi = \sum_{n\in\mathbb{Z}} \hat{\phi}_n e^{\mathrm{i}\alpha_n x_1} \mapsto \left.\frac{\partial v}{\partial\nu}\right|_{\Gamma_+} = \mathrm{i}\sum_{n\in\mathbb{Z}} \beta_n^+ \hat{\phi}_n e^{\mathrm{i}\alpha_n x_1},$$

and the growth bound $|\beta_n^+| \le C|n|$, $n \in \mathbb{Z}$, shows that Λ^+ is bounded from $H_\alpha^s(\Gamma_+)$ to $H_\alpha^{s-1}(\Gamma_+)$ for $s \in \mathbb{R}$. In a similar way one defines a Dirichlet-to-Neumann operator $\Lambda^- : H_\alpha^{1/2}(\Gamma_-) \to H_\alpha^{-1/2}(\Gamma_-)$ on Γ_- taking into account the lower Rayleigh condition (II.6),

$$\Lambda^- : \phi = \sum_{n\in\mathbb{Z}} \hat{\phi}_n e^{\mathrm{i}\alpha_n x_1} \mapsto \mathrm{i} \sum_{n\in\mathbb{Z}} \beta_n^- \hat{\phi}_n e^{\mathrm{i}\alpha_n x_1}.$$

The sign of Λ^- is chosen such that $\Lambda^- v = \partial v/\partial \nu = -\partial v/\partial x_2$ for v which satisfies (II.6). Both operators $\Lambda^\pm$ depend through their coefficients $\beta_n^\pm$ holomorphically on the frequency $\omega \in \mathbb{C} \setminus \mathcal{E}$, see (I.34) for a definition.

LEMMA II.1. *The Dirichlet-to-Neumann operators $\Lambda^\pm$ are holomorphic operator valued functions of the frequency ω in $\mathbb{C} \setminus \mathcal{E}$. Here,*

$$\mathbb{C}_+ = \{z \in \mathbb{C} : \operatorname{Re} z > 0 \textit{ or } \operatorname{Im} z > 0\}$$

and $\mathcal{E}$ was defined in (II.5). *Additionally, for $\omega > 0$ small enough both operators $\Lambda^\pm$ are coercive, i.e., there exists a constant $C > 0$ that does not depend on ω such that, for ψ, $\varphi \in H_\alpha^{1/2}(\Gamma_\pm)$,*

$$-\operatorname{Re}\left(e^{\mathrm{i}\pi/4} \int_{\Gamma_\pm} \overline{\psi} \Lambda^\pm \psi \, \mathrm{d}s \right) \geq C\omega \|\psi\|^2_{H_\alpha^{1/2}(\Gamma_\pm)}.$$

Proof. We only consider Λ^+ in the proof. Holomorphy of Λ^+ follows from the representation

$$\int_{\Gamma_+} \overline{\varphi} \Lambda^+ \psi \, \mathrm{d}s = 2\pi\mathrm{i} \sum_{n\in\mathbb{Z}} \beta_n^+ \hat{\psi}_n \overline{\hat{\varphi}_n} = 2\pi\mathrm{i} \sum_{n\in\mathbb{Z}} \sqrt{\omega^2 \varepsilon_+ - \alpha_n^2}\, \hat{\psi}_n \overline{\hat{\varphi}_n}, \qquad \text{(II.7)}$$

where φ, $\psi \in H_\alpha^{1/2}(\Gamma_+)$ and $\hat{\psi}_n$, $\hat{\varphi}_n$ denote the Fourier coefficients of the periodic functions $\exp(-\mathrm{i}\alpha\cdot)\psi$ and $\exp(-\mathrm{i}\alpha\cdot)\varphi$, respectively. The coefficient $\beta_n^+ = \sqrt{\omega^2\varepsilon_+ - \alpha_n^2}$ is holomorphic in $\omega \in \mathbb{C} \setminus \mathcal{E}$ and the series in (II.7) is absolutely convergent. This implies holomorphy of the series.

In a side computation, we note that $|\omega^2\varepsilon_+ - \alpha_n^2| = \omega^2\, |\varepsilon_+ - (n+\alpha)^2/\omega^2|$. Moreover,

$$\frac{|(n+\alpha)^2/\omega^2 - \varepsilon_+|}{1+n^2} \to \frac{1}{\omega^2} \qquad \text{as } |n| \to \infty.$$

Choosing $\omega_0 > 0$ small enough, the expression $|\varepsilon_+ - \alpha_n^2/\omega^2|$ does never vanish for $\omega \in (0, \omega_0)$; the choice $\omega_0 = \inf_{n\in\mathbb{Z}, n+\alpha\neq 0}[|n+\alpha|/(\sqrt{2\varepsilon_+})]$ is for instance sufficient. Note here that we consider α as a *fixed* parameter of the problem. Hence, there is a constant C independent of ω such that $|(n+\alpha)^2/\omega^2 - \varepsilon_+| \geq C(1+n^2)$ for $0 < \omega < \omega_0$. Using the representation in (II.7) another time, we compute

$$-\operatorname{Re}\left(e^{\mathrm{i}\pi/4}\int_{\Gamma_+}\overline{\psi}\Lambda^+\psi\,\mathrm{d}s\right) = \sqrt{2}\pi\sum_{n\in\mathbb{Z}}\left|\omega^2\varepsilon_+ - \alpha_n^2\right|^{1/2}\left|\hat{\psi}_n\right|^2$$
$$\geq C\omega\sum_{n\in\mathbb{Z}}(1+n^2)^{1/2}\left|\hat{\psi}_n\right|^2 \geq C\omega\|\psi\|^2_{H^{1/2}(\Gamma_+)}$$

for $\omega \in (0,\omega_0)$. □

To obtain a variational formulation of the direct transmission scattering problem, assume that $u^t \in H^1_{\alpha,\mathrm{loc}}(\Pi)$ is a α-quasiperiodic solution that solves the Helmholtz equation in Π and satisfies the boundary conditions (II.2). We multiply the Helmholtz equation (II.1) with $\overline{v}$ for some $v \in H^1_\alpha(D_\pm)$ and apply Green's first identity, yielding

$$\int_{D_\pm}\left(\nabla u^t\nabla\overline{v} - \omega^2\varepsilon_\pm u^t\overline{v}\right)\mathrm{d}x - \int_{\partial D_\pm}\overline{v}\frac{\partial u^t}{\partial\nu_\pm}\,\mathrm{d}s = 0, \qquad \text{(II.8)}$$

where $\nu_\pm$ is the exterior normal to $D_\pm$. The contributions of the integral over $\partial D_\pm$ in (II.8) on the vertical lines cancel due to α-quasiperiodicity of u^t and v. The transmission condition (II.2) implies

$$\int_{D_+\cup D_-}\left(\nabla u^t\nabla\overline{v} - \omega^2\varepsilon u^t\overline{v}\right)\mathrm{d}x = \int_{\Gamma_-}\overline{v}\frac{\partial u^t}{\partial\nu}\,\mathrm{d}s + \int_{\Gamma_+}\overline{v}\frac{\partial u^t}{\partial\nu}\,\mathrm{d}s,$$

where ε is ε_+ in D_+ and ε_- in D_-. Rayleigh's conditions (II.3) and (II.6) give

$$\frac{\partial u^t}{\partial\nu} = \frac{\partial u^i}{\partial\nu} + \frac{\partial u^s}{\partial\nu} = \frac{\partial u^i}{\partial\nu} + \Lambda^\pm u^s$$
$$= \frac{\partial u^i}{\partial\nu} + \Lambda^\pm(u^t - u^i) = \Lambda^\pm u^t + \left(\frac{\partial u^i}{\partial\nu} - \Lambda^\pm u^i\right) \qquad \text{on } \Gamma_\pm.$$

Hence we obtain

$$\mathcal{B}(u^t, v) = \int_{\Gamma_+}\overline{v}\left(\frac{\partial u^i}{\partial\nu} - \Lambda^+ u^i\right)\mathrm{d}s \qquad \text{for all } v \in H^1_\alpha(D), \qquad \text{(II.9)}$$

with a sesquilinear form $\mathcal{B}$ defined by

$$\mathcal{B}(u^t, v) := \int\limits_{D_+ \cup D_-} \left(\nabla u^t \nabla \overline{v} - \omega^2 \varepsilon u^t \overline{v}\right) \mathrm{d}x - \int\limits_{\Gamma_+} \overline{v} \Lambda^+ u^t \,\mathrm{d}s - \int\limits_{\Gamma_-} \overline{v} \Lambda^- u^t \,\mathrm{d}s. \tag{II.10}$$

It is well known that any weak solution in $H^1_{\alpha,\mathrm{loc}}(\Pi)$ of the scattering problem (II.1)-(II.3) solves (II.9) and any solution of the latter formulation can be uniquely extended into Π to a solution of (II.1)-(II.3), see [8].

We briefly state existence and uniqueness results for the variational problem (II.9).

LEMMA II.2 (Theorems 3.1, 3.2 in [23]). *For $\omega > 0$ small enough, the form $\mathcal{B}$ is coercive. Moreover, $\mathcal{B}$ satisfies a Gårding inequality for arbitrary $\omega \in \mathbb{C}_+$.*

By Gårding's inequality and the Fredholm alternative [63, Theorem 2.27], uniqueness of solution of the homogenous problem yields existence and uniqueness for the inhomogeneous problem. One can prove uniqueness of the transmission problem for all frequencies as it is done in [44, Lemma 2], see also [72, Chapter 2]. Therefore, however, additional geometric assumptions on Γ have to be imposed. We do not want to follow this path, but apply the following result, which is known as holomorphic Fredholm theory, see [29, Theorem I.5.1] or [2].

THEOREM II.3. *Assume that $\Omega \subset \mathbb{C}$ is an open connected set and $F_\mu : \mathsf{H} \to \mathsf{H}$ is for all $\mu \in \Omega$ a Fredholm operator on a Hilbert space H. Assume moreover that F_μ is holomorphic in μ. Then for all $\mu \in \Omega$, except possibly for a sequence of isolated points, the equation*

$$F_\mu \varphi = 0$$

has the same number of linearly independent solutions.

Combination of the latter theorem with Lemma II.2 and the observation that the problem is always uniquely solvable for complex frequencies [8] yields the following existence result.

THEOREM II.4. *Let $f \in H^1_\alpha(D)^*$ be a continuous antilinear form on $H^1_\alpha(D)$. For all but an at most countable sequence of positive frequencies $(\omega_j)_{j\in\mathbb{N}} \subset \mathbb{R}_{>0}$ without finite accumulation point, the variational problem* *to find* $u^t \in H^1_\alpha(D)$ *such that*

$$\mathcal{B}(u^t, v) = f(v) \qquad \textit{for all } v \in H^1_\alpha(D)$$

is uniquely solvable.

In the last theorem we have chosen to pay the price of possibly "losing" the discrete set of frequencies $(\omega_j)_{j\in\mathbb{N}}$. We gained not to have to impose that Γ is graph of a function. This choice is motivated by the application of several Dirichlet-to-Neumann operators on Γ to treat the inverse problem later on. For the existence and invertibility of such operators we have to exclude a discrete set of frequencies anyway, since the Neumann problem cannot be shown to be uniquely solvable even if Γ is assumed to be graph of a function.

The above transmission scattering problem (II.9) can be reformulated as a transmission problem for the scattered field u^s, where $u^s = u^t - u^i$ in Ω_+ and $u^s = u^t$ in Ω_-. For simplicity we replace now u^s by u. For $g = -\left.u^i\right|_\Gamma \in H^{1/2}_\alpha(\Gamma)$ and $h = -\left.\partial u^i/\partial\nu\right|_\Gamma \in H^{-1/2}_\alpha(\Gamma)$, u solves

$$\Delta u + \omega^2 \varepsilon u = 0 \quad \text{in } \Omega_+ \cup \Omega_-, \quad [u]_\Gamma = g, \quad \left[\frac{\partial u}{\partial \nu}\right]_\Gamma = h, \qquad \text{(II.11)}$$

and satisfies the Rayleigh conditions (II.3) and (II.6). This problem can of course be posed for general $(g,h) \in H^{1/2}_\alpha(\Gamma) \times H^{-1/2}_\alpha(\Gamma)$. For the variational formulation of (II.11) we transform the jump of the Dirichlet values into the variational formulation using a function $\widetilde{g}$ that satisfies $\widetilde{g}|_\Gamma = g$. Since the trace operator has a continuous right inverse [63, Theorem 3.37], there is $\widetilde{g} \in H^1(D_+)$ with $\widetilde{g}|_\Gamma = g$ such that $\|\widetilde{g}\|_{H^1_\alpha(D_+)} \leq C\|g\|_{H^{1/2}_\alpha(\Gamma)}$ for some constant C independent of g. The problem of finding u, solution of (II.11), is then equivalent to find $v \in H^1_\alpha(D)$, solution of

$$\mathcal{B}(v, w) = \int_\Gamma h\overline{w}\,\mathrm{d}s - \int_{D_+} \left(\nabla\widetilde{g}\nabla\overline{w} - \omega^2\varepsilon_+\widetilde{g}\overline{w}\right)\mathrm{d}x \qquad \text{(II.12)}$$

for all $w \in H^1_\alpha(D)$.

THEOREM II.5. *Problem* (II.11) *is uniquely solvable for all but an at most countable sequence of positive frequencies without finite accumulation point.*

Proof. Theorem II.4 directly implies that the variational problem (II.12) is uniquely solvable for all but an at most countable sequence of frequencies. □

II-3. Quasiperiodic Potentials

Apart from variational formulations one can also apply integral equation methods to solve transmission scattering problems. We will not follow this approach here, but need to introduce the associated potentials and boundary integral operators as well, since they are important for the Factorization method. A basic ingredient for quasiperiodic potentials is the α-quasiperiodic Green's function

$$G(x,y) = \frac{\mathrm{i}}{4\pi} \sum_{n\in\mathbb{Z}} \frac{1}{\beta_n} e^{\mathrm{i}\alpha_n(x_1-y_1)+i\beta_n|x_2-y_2|}, \qquad x \neq y,\ x,y \in \Pi, \quad \text{(II.13)}$$

where we set, by abuse of notation, $\alpha_n = n + \alpha$ and $\beta_n = \sqrt{\omega^2\varepsilon - \alpha_n^2}$ for permittivity ε which is either ε_+ or ε_-. For the choice ε_+, G is the α-quasiperiodic Green's function for wavenumber $\omega\sqrt{\varepsilon_+}$, whereas for the choice ε_- we obtain the α-quasiperiodic Green's function for wavenumber $\omega\sqrt{\varepsilon_-}$. Of course, $\beta_n \neq 0$ for all $n \in \mathbb{N}$ is necessary for G to be well defined.

As $x_2 - y_2 > 0$ for $x_2 > h$ and $y \in \Gamma$ it holds that $|x_2 - y_2| = x_2 - y_2$, thus $G(\cdot, y)$ satisfies the upper Rayleigh condition (II.3): $G(x,y) = \sum g_n(y) e^{\mathrm{i}(\alpha_n x_1 + \beta_n x_2)}$ for $x_2 > h$, $y \in \Gamma$, with coefficients

$$g_n(y) = \frac{\mathrm{i}}{4\pi\beta_n} e^{-\mathrm{i}(\alpha_n y_1 + \beta_n y_2)}, \qquad y \in \Gamma. \quad \text{(II.14)}$$

Concerning the gradient of G, one computes

$$\nabla_y G(x,y) = \sum_{n\in\mathbb{Z}} \nabla g_n(y) e^{\mathrm{i}(\alpha_n x_1 + \beta_n x_2)}, \qquad x_2 > h,\ y \in \Gamma.$$

$G(\cdot, y)$ and $\nabla_y G(\cdot, y)$ also satisfy the lower Rayleigh condition (II.6). These radiation properties carry over to single and double layer potentials defined by

$$\mathrm{SL}\,\phi(x) = \int_\Gamma G(x,y)\phi(y)\,\mathrm{d}s(y), \qquad x \in \Pi \setminus \Gamma, \tag{II.15}$$

$$\mathrm{DL}\,\psi(x) = \int_\Gamma \frac{\partial G}{\partial \nu(y)}(x,y)\psi(y)\,\mathrm{d}s(y), \qquad x \in \Pi \setminus \Gamma. \tag{II.16}$$

The single layer potential gives rise to the operator SL : $H_\alpha^{-1/2}(\Gamma) \to H^1_{\alpha,\mathrm{loc}}(\Pi)$, and the double layer potential gives rise to DL : $H_\alpha^{1/2}(\Gamma) \to H^1_{\alpha,\mathrm{loc}}(\Pi \setminus \Gamma)$. These mapping properties follow from corresponding properties of the usual potential operators, defined using the free space Green's function, because G differs from the free-space Green's function by a smooth function [44]. Jump relations and mapping properties of the potentials are announced in the following theorem, which is proved as in [2, Chapter 3.1].

THEOREM II.6. *The following jump relations hold for* $\phi \in H_\alpha^{-1/2}(\Gamma)$ *and* $\psi \in H_\alpha^{1/2}(\Gamma)$,

$$[\mathrm{SL}\,\phi]_\Gamma = 0, \quad \left[\frac{\partial}{\partial\nu}\mathrm{SL}\,\phi\right]_\Gamma = -\phi, \quad [\mathrm{DL}\,\psi]_\Gamma = \psi, \quad \left[\frac{\partial}{\partial\nu}\mathrm{DL}\,\psi\right]_\Gamma = 0. \tag{II.17}$$

Thus, the operators

$$S : H_\alpha^{-1/2}(\Gamma) \to H_\alpha^{1/2}(\Gamma), \qquad S\phi := \mathrm{SL}\,\phi|_\Gamma, \tag{II.18}$$

$$K : H_\alpha^{1/2}(\Gamma) \to H_\alpha^{1/2}(\Gamma), \qquad K\psi := \mathrm{DL}\,\psi|_\Gamma^+ + \mathrm{DL}\,\psi|_\Gamma^-, \tag{II.19}$$

$$\widetilde{K} : H_\alpha^{-1/2}(\Gamma) \to H_\alpha^{-1/2}(\Gamma), \quad \widetilde{K}\phi := \frac{\partial}{\partial\nu}\mathrm{SL}\,\phi\Big|_\Gamma^+ + \frac{\partial}{\partial\nu}\mathrm{SL}\,\phi\Big|_\Gamma^-, \tag{II.20}$$

$$T : H_\alpha^{1/2}(\Gamma) \to H_\alpha^{-1/2}(\Gamma), \qquad T\psi := -\frac{\partial}{\partial\nu}\mathrm{DL}\,\psi\Big|_\Gamma, \tag{II.21}$$

are well defined and bounded and are related to single and double layer potentials as follows,

$$\mathrm{SL}\,\phi|_\Gamma = S\phi, \qquad \mathrm{DL}\,\psi|_\Gamma^\pm = K\psi \pm \frac{1}{2}\psi,$$

$$\frac{\partial}{\partial\nu}\mathrm{SL}\,\phi\Big|_\Gamma^\pm = \widetilde{K}\phi \mp \frac{1}{2}\phi, \qquad \frac{\partial}{\partial\nu}\mathrm{DL}^\pm\,\psi\Big|_\Gamma = -T\psi.$$

Next we define the α-quasiperiodic Dirichlet Green's function $G_{\rm D}$ for the half space

$$G_{\rm D}(x,y) = G(x,y) - G(x',y), \quad x' = (x_1, -x_2),\, x_1 \geq 0,\, x \neq y. \quad \text{(II.22)}$$

which vanishes on $\{x \in \Pi : x_2 = 0\}$. The Rayleigh coefficients of $G_{\rm D}$ in Ω_+ are easily determined to be $g_n(y) - g_n(y')$. We denote the potential operators defined with the help of $G_{\rm D}$ by $S_{\rm D}$, $K_{\rm D}$, $\widetilde{K}_{\rm D}$ and $T_{\rm D}$.

A helpful fact for our study of the periodic inverse transmission problem is that the single and double layer potentials are holomorphic operator-valued functions of the frequency, see (I.34) for a definition.

THEOREM II.7. *The potentials* SL : $H_\alpha^{-1/2}(\Gamma) \to H_\alpha^1(D)$ *and* DL : $H_\alpha^{1/2}(\Gamma) \to H_\alpha^1(D\backslash\Gamma)$ *depend holomorphically on the frequency* $\omega \in \mathbb{C}_+\backslash\mathcal{E}$.

Proof. We prove the claim for the single layer potential with wavenumber $k = \omega\sqrt{\varepsilon_+}$. The proof for the double layer potential or the corresponding Dirichlet potentials is essentially the same. According to [67] it is sufficient to show that SL depends continuously on the frequency and that for an arbitrary triangle in $\mathbb{C}_+ \setminus \mathcal{E}$ with boundary γ, the integral $\int_\gamma \mathrm{SL}_\omega \phi \,\mathrm{d}\omega$ vanishes in $H_\alpha^1(D)$. Let us first have a look at continuity.

For $\omega_0 \in \mathbb{C}_+ \setminus \mathcal{E}$ and $\omega \in B_r(\omega_0) \subset \mathbb{C}_+ \setminus \mathcal{E}$ with $r > 0$ small enough, we set $w = \mathrm{SL}_\omega \phi$ and $w_0 = \mathrm{SL}_{\omega_0} \phi$ for $\phi \in H_\alpha^{-1/2}(\Gamma)$. The subscripts ω and ω_0 indicate the frequency defining the respective wavenumber of the integral operators. The difference $w - w_0$ solves

$$\Delta(w - w_0) + \omega_0^2 \varepsilon_+ (w - w_0) = \varepsilon_+(\omega_0^2 - \omega^2) w \qquad \text{in } D.$$

and the jump conditions

$$[w - w_0]_\Gamma = \left[\frac{\partial w}{\partial \nu} - \frac{\partial w_0}{\partial \nu}\right]_\Gamma = 0.$$

In case that $\omega_0^2 \varepsilon_+$ is no interior Dirichlet eigenvalue of $-\Delta$ in D with quasiperiodic boundary conditions on the left and right boundary, standard results on well-posedness of elliptic problems [63, Chapter 4] imply

$$\begin{aligned}\|w - w_0\|_{H_\alpha^1(D)} &\leq C\varepsilon_+ |\omega - \omega_0| \, \|w\|_{L^2(D)} \\ &\quad + C\|w - w_0\|_{H^{1/2}(\Gamma_+)} + C\|w - w_0\|_{H^{1/2}(\Gamma_-)}. \end{aligned} \quad \text{(II.23)}$$

However, if $\omega_0^2\varepsilon_+$ is a Dirichlet eigenvalue, by Holmgren's lemma $\omega_0^2\varepsilon_+$ cannot be a Neumann eigenvalue with quasiperiodic boundary conditions on the left and right boundary of D and hence we would obtain a similar estimate relying on the Neumann boundary values of $w - w_0$ on $\Gamma_\pm$, which then allows to proceed the proof in an analogous fashion as we do it now.

Since $\|w\|_{L^2(D)}$ is uniformly bounded for all $\omega \in B_r(\omega_0)$ by the norm of ϕ, it merely remains to show that $\|w - w_0\|_{H^{1/2}(\Gamma_\pm)}$ tends to zero as $\omega_0 \to \omega$, uniformly in ϕ. We only need to show this for Γ_+, the proof for Γ_- is basically the same. Since

$$(w - w_0)|_{\Gamma_+} = \int_\Gamma (G_\omega(\cdot, y) - G_{\omega_0}(\cdot, y))\, \phi(y)\, \mathrm{d}s(y)\bigg|_{\Gamma_-}, \tag{II.24}$$

we need to estimate an integral operator with smooth kernel $G_\omega(\cdot, y) - G_{\omega_0}(\cdot, y)$. Note that the series

$$G_\omega(x, y) = \frac{\mathrm{i}}{4\pi} \sum_{n\in\mathbb{Z}} \frac{1}{\beta_n(\omega)} e^{\mathrm{i}\alpha_n(x_1-y_1)+i\beta_n(\omega)|x_2-y_2|},$$

converges absolutely and uniformly for $x \in \Gamma_+$ and $y \in \Gamma$, and the summands of the series for frequency ω converge to those of the series for frequency ω_0 as ω tends to ω_0, uniformly in x and y. Hence, there is $\delta : [0, \infty) \to [0, \infty)$ with $\delta(s) \to 0$ as $s \to 0$ such that

$$|G_\omega(x, y) - G_{\omega_0}(x, y)| \le \delta(|\omega - \omega_0|)$$

for $x \in \Gamma_+$ and $y \in \Gamma$. Using the same technique, one shows that for any multiindex $\boldsymbol{\alpha} \in \mathbb{N}^2$ and $\boldsymbol{\beta} \in \mathbb{N}^2$ there is $\delta = \delta(\boldsymbol{\alpha}, \boldsymbol{\beta}) : [0, \infty) \to [0, \infty)$ with $\delta(s) \to 0$ as $s \to 0$ such that

$$\left| \frac{\partial^{\boldsymbol{\alpha}}}{\partial x^{\boldsymbol{\alpha}}} \frac{\partial^{\boldsymbol{\beta}}}{\partial y^{\boldsymbol{\beta}}} (G_\omega(x, y) - G_{\omega_0}(x, y)) \right| \le \delta(|\omega - \omega_0|)$$

for $x \in \Gamma_+$ and $y \in \Gamma$. In consequence, (II.24) shows that there is a function δ as above such that $\|w - w_0\|_{H_\alpha^{1/2}(\Gamma_+)} \le C\|w - w_0\|_{H_\alpha^1(\Gamma_+)} \le C\delta(|\omega - \omega_0|)\|\phi\|_{H_\alpha^{-1/2}(\Gamma)}$. Therefore we conclude from the estimate

$$\|w - w_0\|_{H_\alpha^1(D)} \le C\varepsilon_+|\omega - \omega_0|\, \|w\|_{L^2(D)} + C\delta(|\omega - \omega_0|)\|\phi\|_{H_\alpha^{-1/2}(\Gamma)}$$

that SL depends continuously on the frequency.

Next we prove that $\int_\gamma \mathrm{SL}_\omega\,\phi\,\mathrm{d}\omega = 0$ for all triangles in $\mathbb{C}_+ \setminus \mathcal{E}$ with boundary γ. Note that the integral over ω is Banach space valued (since SL_ω depends continuously on ω, we can for instance use the operator-valued Riemann integral here, see, e.g., [61, Appendix A]). Then

$$\left[\int_\gamma \mathrm{SL}_\omega\,\phi\,\mathrm{d}\omega\right](x) = \int_\gamma \langle G_\omega(x,\cdot),\,\phi\rangle_{L^2(\Gamma)}\,\mathrm{d}\omega$$
$$= \langle \textstyle\int_\gamma G_\omega(x,\cdot)\,\mathrm{d}\omega,\,\phi\rangle_{L^2(\Gamma)}$$

for $x \notin \Gamma$. Interchanging the integrals for fixed x is justified by Fubini's theorem for smooth ϕ and for arbitrary $\phi \in H_\alpha^{-1/2}(\Gamma)$ since $\int_\gamma \cdot\,\mathrm{d}\omega$ is a continuous operation on functions depending continuously on ω. It remains to prove that $\int_\gamma G_\omega(x,y)\,\mathrm{d}\omega = 0$ for $y \in \Gamma$ and $x \neq y$.

Since $G(\cdot,y)$ solves the Helmholtz equation in $\Pi \setminus \{y\}$ it is real analytic. In consequence, $v_y := \int_\gamma G_\omega(\cdot,y)\,\mathrm{d}\omega$ is also real analytic. Hence, if v_y vanishes on $\{x \in \Pi : x_2 > h\}$, then v_y vanishes entirely in Ω_+. For $x_2 > h$ and $y \in \Gamma$ the Dirichlet Green's function is given by the uniformly convergent series

$$G_\omega(x,y) = \frac{i}{4\pi}\sum_{n\in\mathbb{Z}} \frac{1}{\beta_n} e^{\mathrm{i}(\alpha_n(x_1-y_1)+\beta_n(x_2-y_2))},$$

which depends on ω through β_n, and we find

$$v_y(x) = \frac{i}{4\pi}\sum_{n\in\mathbb{Z}} e^{\mathrm{i}\alpha_n(x_1-y_1)} \int_\gamma \frac{1}{\beta_n} e^{\mathrm{i}\beta_n(x_2-y_2)}\,\mathrm{d}\omega \qquad \text{in } \Omega_+.$$

The last integral vanishes since the integrand

$$\omega \mapsto \frac{1}{\beta_n} e^{\mathrm{i}\beta_n(x_2-y_2)} = \frac{1}{\sqrt{\omega^2\varepsilon_- - \alpha_n^2}} e^{\mathrm{i}\sqrt{\omega^2\varepsilon_- - \alpha_n^2}(x_2-y_2)}$$

is a holomorphic function of ω in $\mathbb{C}_+ \setminus \mathcal{E}$. Since $G_\omega(\cdot,y)$ also satisfies the lower Rayleigh condition (II.6) in Ω_- the same argument shows that v_y vanishes entirely in $\Pi \setminus \{y\}$. Therefore $\int_\gamma G_\omega(\cdot,y)\,\mathrm{d}\omega = 0$ in $H_\alpha^{1/2}(\Gamma)$ for all $y \in \Gamma$ and $x \in \Omega_+$. $\square$

Since the trace operator and the normal derivative are independent of the frequency ω, the following corollary holds.

Corollary II.8. *The potential operators S, K, $\widetilde{K}$ and T as well as the correspondent operators defined via the Dirichlet Green's function G_{D} are holomorphic operator valued functions of the frequency $\omega \in \mathbb{C}_+ \setminus \mathcal{E}$.*

II-4. Auxiliary Scattering Problems

The analysis of the Factorization method for the inverse transmission problem requires solvability of several auxiliary boundary value problems above and below Γ. The related solution operators are necessary to establish existence of certain Dirichlet-to-Neumann and Neumann-to-Dirichlet operators. It becomes later on clear that we additionally need to solve these problems for both material parameters $\varepsilon_\pm$ above and below Γ. For $g \in H_\alpha^{1/2}(\Gamma)$ we consider the problem of finding $u \in H^1_{\alpha,\mathrm{loc}}(\Omega)$ that satisfies

$$\Delta u + \omega^2 \varepsilon_\pm u = 0 \quad \text{in } \Omega_+, \qquad u = g \quad \text{on } \Gamma, \tag{II.25}$$

and the upper Rayleigh condition (II.3). The boundary data g can also be used to pose a Dirichlet boundary value problem *below* Γ: find $u \in H^1_{\alpha,\mathrm{loc}}(\Omega_-)$ that fulfills

$$\Delta u + \omega^2 \varepsilon_\pm u = 0 \quad \text{in } \Omega_-, \qquad u = g \quad \text{on } \Gamma, \tag{II.26}$$

and the Rayleigh condition (II.6). The Neumann problems of interest are the following. For $h \in H_\alpha^{-1/2}(\Gamma)$ we want to find $u \in H^1_{\alpha,\mathrm{loc}}(\Omega_+)$ that satisfies

$$\Delta u + \omega^2 \varepsilon_\pm u = 0 \quad \text{in } \Omega_+, \qquad \frac{\partial u}{\partial \nu} = h \quad \text{on } \Gamma, \tag{II.27}$$

and the Rayleigh condition (II.3). As for the Dirichlet case we also take the boundary values h to pose the Neumann problem in the lower domain,

$$\Delta u + \omega^2 \varepsilon_\pm u = 0 \quad \text{in } \Omega_-, \qquad \frac{\partial u}{\partial \nu} = h \quad \text{on } \Gamma, \tag{II.28}$$

and condition (II.6). Again, holomorphic Fredholm theory implies that all Dirichlet and Neumann problems are uniquely solvable for all but a countable sequence of real frequencies. For simplicity we define the set of exceptional frequencies $\mathcal{E}_{\mathrm{d}}$ to be the union of $\mathcal{E}$ with all frequencies such that one of the Dirichlet, Neumann or transmission problems fails to be uniquely solvable. Frequencies $\omega \notin \mathcal{E}_\mathrm{d}$ are called *regular*. For such ω we can define Dirichlet-to-Neumann operators $\Upsilon_{(\varepsilon_\pm,\pm)}$ on $H^{1/2}_\alpha(\Gamma)$ that are invertible. Indeed, define for $g \in H^{1/2}_\alpha(\Gamma)$ the distribution $\Upsilon_{(\varepsilon_\pm,+)}g$ to be the Neumann boundary values of the solution u of the auxiliary problem (II.25),

$$\Upsilon_{(\varepsilon_\pm,+)}g = \frac{\partial u}{\partial \nu} \in H^{-1/2}_\alpha(\Gamma). \tag{II.29}$$

As we assumed ω to be a regular frequency, the Neumann problem (II.27) is uniquely solvable for the boundary data $\Upsilon_{(\varepsilon_\pm,+)}g$ and we conclude that $\Upsilon_{(\varepsilon_\pm,+)}$ is invertible since it is bounded and bijective. In a similar way we define $\Upsilon_{(\varepsilon_\pm,-)}$ for $\omega \notin \mathcal{E}_\mathrm{d}$ as the Dirichlet-to-Neumann operator that maps $g \in H^{1/2}_\alpha(\Gamma)$ to the Neumann boundary values of the solution of (II.26). In Lemma II.1 it was established that $\Lambda^\pm$ is holomorphic in ω in $\mathbb{C}_+ \setminus \mathcal{E}$. The same holds for $\Upsilon_{(\varepsilon_\pm,\pm)}$ if we replace $\mathcal{E}$ by $\mathcal{E}_\mathrm{d}$.

THEOREM II.9. *The Neumann-to-Dirichlet operator $\Upsilon_{(\varepsilon_\pm,\pm)}$ as well as its inverse is holomorphic in ω in $\mathbb{C}_+\setminus\mathcal{E}_\mathrm{d}$ and coercive for small frequencies $\omega > 0$.*

Proof. We only prove the claim for $\Upsilon = \Upsilon_{(\varepsilon_+,+)}$.

Since by definition $\Upsilon\psi = \partial u/\partial\nu$ for u which solves (II.25) for the boundary values $u = \psi$, it is sufficient to show that u is a holomorphic function of ω in $H^1_\alpha(D_+)$. Our assumption that $\omega \notin \mathcal{E}_\mathrm{d}$ implies that the single layer operator S on Γ for the wavenumber $\omega\varepsilon_+^{1/2}$ is an isomorphism, since it is a Fredholm operator anyway [7]. Therefore we can represent $u \in H^1_\alpha(D_+)$ in the form $u = \mathrm{SL}\,\phi$ where $\phi \in H^{-1/2}_\alpha(\Gamma_+)$ solves the boundary integral equation $S\phi = \psi$. Equivalently, $\phi = S^{-1}\psi$. Corollary (II.8) implies that S^{-1} is holomorphic in the frequency ω. Therefore ϕ also depends holomorphically on the frequency. The equation $u = \mathrm{SL}\,\phi$ implies that u is holomorphic in ω, because the composition of two holomorphic functions is holomorphic.

To show coercivity of Υ, assume that $\omega > 0$ is so small that Lemma II.1 applies. We use Green's first identity (note that the normal ν on Γ was defined to point into D_+) to find

$$-\int_\Gamma \overline{\psi}\Upsilon\psi \,\mathrm{d}s = \int_{D_+} \left(|\nabla u|^2 - \omega^2\varepsilon_+|u|^2\right)\mathrm{d}x - \int_{\Gamma_+} \overline{u}\Lambda^+ u \,\mathrm{d}s$$

and conclude by Lemma II.1 that

$$-\mathrm{Re}\left(e^{\mathrm{i}\pi/4}\int_\Gamma \overline{\psi}\Upsilon\psi\,\mathrm{d}s\right) \geq \sqrt{2}\|\nabla u\|^2_{L^2(D_+)} + C\omega\,\|u\|^2_{H^{1/2}_\alpha(\Gamma_+)} - \sqrt{2}\omega^2\varepsilon_+\|u\|^2_{L^2(D_+)}$$
$$\geq C\omega\|u\|^2_{H^1_\alpha(D)} - \sqrt{2}\omega^2\varepsilon_+\|u\|^2_{L^2(D_+)},$$

where we used that $\|\nabla u\|_{L^2(D_+)} + \|u\|_{H^{1/2}_\alpha(\Gamma_+)}$ is an equivalent norm on $H^1_\alpha(D_+)$, see [23, Section 3.1]. As ω^2 tends to zero faster than ω we obtain coercivity of Υ for small frequencies $\omega > 0$. □

A scattered field u that solves the transmission problem (II.11) is due to the Rayleigh expansion condition (II.3) completely characterized in $\{x \in \Pi : x_2 > h\}$ by the Fourier coefficients $(u_n^+)_{n\in\mathbb{Z}}$ of $\exp(-\mathrm{i}\alpha\cdot)u|_{\Gamma_+}$. This sequence is referred to as the *Rayleigh sequence* of u. We define the *data-to-pattern operator* L to map a pair of transmission values (g,h) to the upper Rayleigh sequence of the solution u of (II.11),

$$L: H^{1/2}_\alpha(\Gamma)\times H^{-1/2}_\alpha(\Gamma) \to \ell^2, \qquad (g,h)\mapsto (u_n^+)_{n\in\mathbb{Z}}. \tag{II.30}$$

The following proposition shows that the Dirichlet-to-Neumann operator $\Upsilon_{(\varepsilon_-,-)}$ characterizes the kernel of L.

PROPOSITION II.10. *Let $\omega \in \mathbb{C}_+ \setminus \mathcal{E}_\mathrm{d}$ be a regular frequency. Then the data-to-pattern operator $L: H^{1/2}_\alpha(\Gamma)\times H^{-1/2}_\alpha(\Gamma) \to \ell^2$ is compact and its range is dense in ℓ^2. The kernel of L is given by*

$$\ker L := \{(g, \Upsilon_{(\varepsilon_-,-)}g) : \ g \in H^{1/2}_\alpha(\Gamma)\}.$$

Proof. Compactness of L follows from smoothing properties of the Helmholtz equation. Indeed, suppose that $L(g,h) = (u_n)$ for $(g,h) \in H^{1/2}_\alpha(\Gamma)\times H^{-1/2}_\alpha(\Gamma)$ and that u denotes the solution of (II.11). By well known regularity results [22], $\exp(-\mathrm{i}\alpha\cdot)u$ is a smooth periodic function for $x_2 \geq h$. The Fourier coefficients u_n^+ of $\exp(-\mathrm{i}\alpha\cdot)u|_{\Gamma_+}$ exhibit hence super-algebraic decay. In particular, the sequence (u_n^+) belongs to the weighted sequence space h^s, $s\in\mathbb{R}$, with norm $\|(a_n)\|^2_{h^s} = \sum_{n\in\mathbb{Z}}(1+n^2)^s|a_n|^2$, for arbitrary $s > 0$. The sequence space h^s is isomorphic to the Sobolev space

$H^s_{\text{per}}(\Gamma_+)$ with periodic boundary conditions, because

$$I : (a_n) \mapsto \sum_{n\in\mathbb{Z}} a_n e^{\mathrm{i}n(\cdot)} \tag{II.31}$$

defines a norm isomorphism. The well known compactness of the embedding $H^s_{\text{per}}(\Gamma_+) \hookrightarrow H^0_{\text{per}}(\Gamma_+) = L^2(\Gamma_+)$, $s > 0$ yields compactness of $h^s \hookrightarrow h^0 = \ell^2$, especially, L is compact.

To prove denseness of the range of L one shows that all sequences $e_j := (0, \dots, 0, 1, 0, \dots)$, where the jth entry equals 1, belong to the range of L. We observe that $u = \exp(\mathrm{i}(\alpha_n x_1 + \beta_n x_2))$ in Ω_+, $u = 0$ in Ω_-, solves (II.11) for the data

$$\left[e^{\mathrm{i}(\alpha_n x_1 + \beta_n^+ x_2)}\Big|_\Gamma, \frac{\partial}{\partial\nu} e^{\mathrm{i}(\alpha_n x_1 + \beta_n^+ x_2)}\Big|_\Gamma \right] \in H^{1/2}_\alpha(\Gamma) \times H^{-1/2}_\alpha(\Gamma).$$

and has the Rayleigh sequence e_j.

By our assumption on ω, for $g \in H^{1/2}_\alpha(\Gamma)$ there exists $v \in H^1_\alpha(D_-)$ that solves (II.26) for the Dirichlet data g such that $\partial v/\partial\nu = \Upsilon_{(\varepsilon_-,-)} g$ on Γ. Then u, defined by $u = 0$ in Ω_+, $u = v$ in Ω_-, solves (II.11) for $(g, h) = (g, \Upsilon_{(\varepsilon_-,-)} g)$. Since u vanishes in Ω_+ it is clear that $L(g, \Upsilon_{(\varepsilon_-,-)} g) = 0$. It remains to show that the elements $(g, \Upsilon_{(\varepsilon_-,-)} g)$ form the entire kernel of L. Assume that $L(g, h) = 0$ for some $h \in H^{-1/2}_\alpha(\Gamma)$ and $h \neq \Upsilon_{(\varepsilon_-,-)} g$. By linearity, $L(0, \Upsilon_{(\varepsilon_-,-)} g - h) = 0$. The associated solution v of the scattering problem (II.11) for the data $(0, \Upsilon_{(\varepsilon_-,-)} g - h)$ vanishes in Ω_+ by analytic continuation. Since the jump of the Dirichlet trace of u on Γ vanishes, u solves (II.26) for $g = 0$. We conclude from the unique solvability of this problem that $u = 0$ below Γ, thus, in all of Π. In consequence, $\Upsilon_{(\varepsilon_-,-)} g = h$ which is a contradiction to our assumption $h \neq \Upsilon_{(\varepsilon_-,-)} g$. □

The interface Γ can be explicitly characterized by the range of L using Rayleigh sequences $(g_n(y))_n$ of the quasiperiodic Green's function, see (II.14).

LEMMA II.11. *The Rayleigh sequence $(g_n(y))_{n\in\mathbb{Z}}$ belongs to the range of L if and only if $y \in \Omega_-$.*

The proof is standard and can in a related setting be found in [3, Lemma 2.2]. See also Lemma I.5 for the related result for the inverse medium problem which is proved in the same way. The crucial ingredient for the proof is the connectedness of Ω_+. We remark further that the claim of Lemma II.11 holds also for the Rayleigh sequence $(g_n(y) - g_n(y'))_{n\in\mathbb{Z}}$ of the Dirichlet Green's function $G_{\mathrm{D}}(\cdot, y)$.

II-5. The Near Field Operator and its Factorization

Let us now turn to the inverse transmission problem for periodic structures. Roughly speaking, it consists in determining the interface Γ from measurements of scattered waves caused by incident plane waves; see page 64 for a precise statement. We restrict ourselves here to the case where we send plane waves to the interface merely from above and measure the scattered waves also only above the interface Γ. Our technique to solve the inverse scattering problem, the Factorization method, is hence applicable even if one has merely access to the object under investigation from one side.

To generate the scattering data for the Factorization method we consider the α-quasiperiodic incident fields

$$\eta_j(x) := \frac{\mathrm{i}}{4\pi\beta_j^+}\left(e^{\mathrm{i}(\alpha_j x_1 - \beta_j^+ x_2)} - e^{\mathrm{i}(\alpha_j x_1 + \beta_j^+ x_2)}\right), \qquad x \in \Omega_+,\ j \in \mathbb{Z}. \tag{II.32}$$

These incident waves are composed by two fields $\exp(\mathrm{i}(\alpha_j x_1 - \beta_j^+ x_2))$ and $\exp(\mathrm{i}(\alpha_j x_1 + \beta_j^+ x_2))$ which are downward and upward propagating, respectively. The upward propagating part has no physical meaning but from an abstract mathematical viewpoint it poses no difficulty: the corresponding total field vanishes entirely. The choice (II.32) for the incident fields allows to analyze the Factorization method for propagating incident fields and has been proposed and discussed in [3].

If one combines several incident fields, the resulting scattered field is by linearity found by correspondingly combined scattered fields. We achieve such linear combinations through sequences $(a_n)_{n\in\mathbb{Z}} \in \ell^2$ and the corresponding operator is denoted by

$$H(a_n) = \left[\sum_{j\in\mathbb{Z}} a_j\, \eta_j|_\Gamma,\ \sum_{j\in\mathbb{Z}} a_j\, \frac{\partial \eta_j}{\partial \nu}\bigg|_\Gamma\right] \in H_\alpha^{1/2}(\Gamma) \times H_\alpha^{-1/2}(\Gamma). \tag{II.33}$$

This operator is the analogue to the Herglotz operator from Section I-3. In the next lemma, we prove that H is a bounded operator on ℓ^2, using the representation of its (yet formal) adjoint. We recall that the Rayleigh sequence $(g_j(x))_{j\in\mathbb{Z}}$ of the Green's function G has been defined in (II.14).

LEMMA II.12. *Let $\omega \notin \mathcal{E}_{\mathrm{d}}$ be a positive regular frequency. Then $H : \ell^2 \to H_\alpha^{1/2}(\Gamma) \times H_\alpha^{-1/2}(\Gamma)$ is bounded, even compact, and its adjoint H^* is given by*

$$H^*(\phi,\psi)$$
$$= \left(\int_\Gamma \left[(g_j(x) - g_j(x'))\phi(x) + \frac{\partial(g_j(x) - g_j(x'))}{\partial\nu}\psi(x)\right] \mathrm{d}s(x)\right)_{j\in\mathbb{Z}}$$

for $(\phi,\psi) \in H_\alpha^{-1/2}(\Gamma) \times H_\alpha^{1/2}(\Gamma)$.

Proof. Denote the two components of $H(a_n)$ by $(H_1(a_n), H_2(a_n))$ and let $(\phi,\psi) \in H_\alpha^{-1/2}(\Gamma) \times H_\alpha^{1/2}(\Gamma)$. Then $\langle H(a_n),(\phi,\psi)\rangle = \langle H_1(a_n),\phi\rangle + \langle H_2(a_n),\psi\rangle$. For the first summand we find

$$\langle H_1(a_n),\phi\rangle = \int_\Gamma \sum_{j\in\mathbb{Z}} a_j\eta_j\overline{\phi}\,\mathrm{d}s = \sum_{j\in\mathbb{Z}} a_j \int_\Gamma \eta_j\overline{\phi}\,\mathrm{d}s = \left\langle a_j, \int_\Gamma \overline{\eta_j}\phi\,\mathrm{d}s\right\rangle_{\ell^2}.$$

The second summand is computed in the same way. Observe that $\overline{\alpha_j} = \alpha_j$ and $\overline{\beta_j^+} = \beta_j^+$ if $\alpha_j^2 \le \omega^2\varepsilon_+$ but $\overline{\beta_j^+} = -\beta_j^+$ else. Therefore

$$\overline{\eta_j}(x) = \begin{cases} \frac{-\mathrm{i}}{4\pi\beta_j^+}\left[e^{-\mathrm{i}(\alpha_j x_1 - \beta_j^+ x_2)} - e^{-\mathrm{i}(\alpha_j x_1 + \beta_j^+ x_2)}\right], & \alpha_j^2 \le \omega^2\varepsilon_+, \\ \frac{i}{4\pi\beta_j^+}\left[e^{-\mathrm{i}(\alpha_j x_1 + \beta_j^+ x_2)} - e^{-\mathrm{i}(\alpha_j x_1 - \beta_j^+ x_2)}\right], & \text{else,} \end{cases}$$
$$= g_j(x) - g_j(x').$$

Concerning boundedness of H, we note that $H^*(\phi,\psi)$ is the Rayleigh sequence of $v = \mathrm{SL_D}\,\phi + \mathrm{DL_D}\,\psi$. The mapping $(\phi,\psi) \mapsto v|_{\Gamma_+}$ is continuous from $H_\alpha^{-1/2}(\Gamma) \times H_\alpha^{1/2}(\Gamma)$ into $H_\alpha^{1/2}(\Gamma_+)$. According to (II.31),

$$I_\alpha : (a_n) \mapsto e^{\mathrm{i}\alpha(\cdot)} \sum_{n\in\mathbb{Z}} a_n e^{\mathrm{i}n(\cdot)} = \sum_{n\in\mathbb{Z}} a_n e^{\mathrm{i}\alpha_n(\cdot)},$$

is an isomorphism between h^s and $H_\alpha^s(\Gamma_+)$. Hence, v is continuously mapped to its Rayleigh coefficients $(v_n) \in h^{1/2} \hookrightarrow \ell^2$, which are just the Fourier coefficients of $e^{-\mathrm{i}\alpha(\cdot)}\, v|_{\Gamma_+}$. Therefore H^* is continuous, even compact, and its adjoint H as well. □

The investigation of the adjoint H^* is also one of the key ideas of the Factorization method. More precisely, the method links the adjoint H^* to some boundary integral operator on Γ.

In the remainder of this section we define and factorize the data operator of the inverse transmission problem. Since this data operator corresponds to measurements of scattered fields in a finite distance away from the interface Γ, it is usually referred to as the near field operator. By definition, the near field operator $N : \ell^2 \to \ell^2$ maps a sequence (a_n) to the upper Rayleigh sequence of the scattered field caused by the incident field $H(a_n)$,

$$N := -LH. \tag{II.34}$$

Equivalently, we could also define $N(a_n)$ via the Fourier coefficients of the restriction $u|_{\Gamma_+}$ (multiplied with $\exp(-\mathrm{i}\alpha\cdot)$) of the scattered field u corresponding to the incident field $H(a_n)$. Such a definition illustrates the meaning of the expression "near field" much better. Obviously, the definition (II.34) is mathematically simpler and therefore we prefer it here. More important, note that N involves data merely measured *above* Γ.

The *inverse transmission problem* is now to reconstruct the interface Γ when given the near field operator N. We solve this problem with the help of the Factorization method. For uniqueness results for a related identification problem, we refer to [24]: In this work the authors show that for small wavenumbers the interface Γ is uniquely determined by the scattered field corresponding to one incident plane wave. In general, several incident fields are needed to obtain scattering data which characterize the interface uniquely. However, the results in [24] rely on measurements of the fields on both sides of the structure. We are not aware of uniqueness results for one sided measurements in the literature, but remark that the Factorization which we develop in this chapter will provide such a uniqueness result.

In the first step we construct an operator $M_{\mathrm{D}} : H_\alpha^{-1/2}(\Gamma) \times H_\alpha^{1/2}(\Gamma) \to H_\alpha^{1/2}(\Gamma) \times H_\alpha^{-1/2}(\Gamma)$ such that the factorization $N = -LM_{\mathrm{D}}^* L^*$ holds. As we have seen in the proof of Lemma II.12, it holds that $H^*(\phi,\psi)$ is the Rayleigh sequence of $v = \mathrm{SL}_{\mathrm{D}}\,\phi + \mathrm{DL}_{\mathrm{D}}\,\psi \in H^1_{\alpha,\mathrm{loc}}(\Omega_+)$. To avoid confusion, we denote the dependence of potential operators with wavenumber

$\omega\sqrt{\varepsilon_\pm}$ on the permittivity $\varepsilon_\pm$ in the following explicitely whenever this is necessary, for instance as follows, $v = \mathrm{SL}_{\mathrm{D}}^{\varepsilon_+}\,\phi + \mathrm{DL}_{\mathrm{D}}^{\varepsilon_+}\,\psi$. Searching for M_{D} such that $H^* = LM_{\mathrm{D}}$ we need to consider the jump of a suitable extension of v to all of $\Pi \setminus \Gamma$ that solves the transmission problem (II.11). One possible choice is

$$v = \begin{cases} \mathrm{SL}_{\mathrm{D}}^{\varepsilon_+}\,\phi + \mathrm{DL}_{\mathrm{D}}^{\varepsilon_+}\,\psi, & x \in \Omega_+, \\ -\,\mathrm{SL}^{\varepsilon_-}\,\phi - \mathrm{DL}^{\varepsilon_-}\,\psi, & x \in \Omega_-. \end{cases} \tag{II.35}$$

This choice of extension is of course not unique and by Lemma II.10 we could take any downward radiating function in Ω_-. The jump of v in (II.35) can be computed by Theorem II.6 and we define the operator M_{D} to map (ϕ, ψ) to this jump,

$$\begin{aligned} M_{\mathrm{D}} : H^{-1/2}(\Gamma) \times H^{1/2}(\Gamma) &\to H^{1/2}(\Gamma) \times H^{-1/2}(\Gamma), \\ (\phi, \psi) &\mapsto \Big(\big[S_{\mathrm{D}}^{\varepsilon_+} + S^{\varepsilon_-}\big]\,\phi + \big[K_{\mathrm{D}}^{\varepsilon_+} + K^{\varepsilon_-}\big]\,\psi, \\ &\qquad \Big[\widetilde{K}_{\mathrm{D}}^{\varepsilon_+} + \widetilde{K}^{\varepsilon_-}\Big]\,\phi - \big[T_{\mathrm{D}}^{\varepsilon_+} + T^{\varepsilon_-}\big]\,\psi \Big). \end{aligned} \tag{II.36}$$

Representing M_{D} in matrix form yields

$$M_{\mathrm{D}} = \begin{pmatrix} S_{\mathrm{D}}^{\varepsilon_+} + S^{\varepsilon_-} & K_{\mathrm{D}}^{\varepsilon_+} + K^{\varepsilon_-} \\ \widetilde{K}_{\mathrm{D}}^{\varepsilon_+} + \widetilde{K}^{\varepsilon_-} & -(T_{\mathrm{D}}^{\varepsilon_+} + T^{\varepsilon_-}) \end{pmatrix}.$$

With this choice of M_{D} we obtain $H^* = LM_{\mathrm{D}}$ and the factorization $N = -LM_{\mathrm{D}}^* L^*$ holds. The middle operator of this factorization is a Fredholm operator.

PROPOSITION II.13. *The middle operator $M_{\mathrm{D}}^* : H_\alpha^{-1/2}(\Gamma) \oplus H_\alpha^{1/2}(\Gamma) \to H_\alpha^{1/2}(\Gamma) \oplus H_\alpha^{-1/2}(\Gamma)$ is a Fredholm operator of index zero.*

Proof. As an auxiliary result, let us first show that for $\omega = \mathrm{i}$, the operators $S = S_{\mathrm{i}}$ and $T = T_{\mathrm{i}}$ are coercive on $H_\alpha^{-1/2}(\Gamma)$ and $H_\alpha^{1/2}(\Gamma)$, respectively, for both permittivities $\varepsilon_\pm$. If we set $u = \mathrm{SL}_{\mathrm{i}}\,\phi$ for $\phi \in H_\alpha^{-1/2}(\Gamma)$, an integration by parts shows that

$$\begin{aligned}\int_\Gamma \overline{\phi} S_{\mathrm{i}} \phi \,\mathrm{d}s &= \int_\Gamma u \left(\frac{\partial}{\partial \nu} \overline{u}\Big|_\Gamma^- - \frac{\partial}{\partial \nu} \overline{u}\Big|_\Gamma^+ \right) \mathrm{d}s \\ &= \int_{D_+ \cup D_-} |\nabla u|^2 + \varepsilon(x) |u|^2 \,\mathrm{d}x - \int_{\Gamma_+} \overline{u} \Lambda^+ u \,\mathrm{d}s - \int_{\Gamma_-} \overline{u} \Lambda^- u \,\mathrm{d}s \\ &\geq C \|u\|_{H^1_\alpha(D_+ \cup D_-)} \geq C \|\phi\|_{H_\alpha^{-1/2}(\Gamma)}.\end{aligned}$$

Here we exploited that for $\omega = \mathrm{i}$ it holds

$$-\int_{\Gamma_\pm} \overline{u} \Lambda^+ u \,\mathrm{d}s = \sum_{n \in \mathbb{Z}} \sqrt{\varepsilon_\pm + \alpha_n^2}\, |\hat{u}_n|^2 \geq C \|u\|^2_{H_\alpha^{1/2}(\Gamma_\pm)}.$$

The proof for the normal derivative of the double layer potential can be shown by an analogous integration by parts.

Especially, for $\omega = \mathrm{i}$, both S_{i} and T_{i} are isomorphisms and hence Fredholm operators of index zero. It is moreover well known that our smoothness assumption on Γ to be of class $C^{3,\alpha}$ implies that the difference of two layer potential operators for different wavenumbers is compact. For a proof we refer to [43] and to [2, Section 3.4] for the case of quasiperiodic potentials. Consequently, S and T are Fredholm operators for all $\omega \in \mathbb{C}$. We already mentioned in Section II-3 that the corresponding Dirichlet layer potential operators are compact perturbations of the usual quasiperiodic ones. We conclude that

$$\begin{aligned}\begin{pmatrix} S_{\mathrm{D}}^{\varepsilon_+} + S^{\varepsilon_-} & 0 \\ 0 & -T_{\mathrm{D}}^{\varepsilon_+} - T^{\varepsilon_-} \end{pmatrix} &= \begin{pmatrix} S_{\mathrm{i}}^{\varepsilon_+} + S_{\mathrm{i}}^{\varepsilon_-} & 0 \\ 0 & -T_{\mathrm{i}}^{\varepsilon_+} - T_{\mathrm{i}}^{\varepsilon_-} \end{pmatrix} \\ &+ \begin{pmatrix} S_{\mathrm{D}}^{\varepsilon_+} - S_{\mathrm{i}}^{\varepsilon_+} & 0 \\ 0 & T_{\mathrm{i}}^{\varepsilon_+} - T_{\mathrm{D}}^{\varepsilon_+} \end{pmatrix} + \begin{pmatrix} S^{\varepsilon_-} - S_{\mathrm{i}}^{\varepsilon_-} & 0 \\ 0 & T_{\mathrm{i}}^{\varepsilon_-} - T^{\varepsilon_-} \end{pmatrix}\end{aligned}$$

is Fredholm of index zero from $H_\alpha^{-1/2}(\Gamma) \oplus H_\alpha^{1/2}(\Gamma)$ into $H_\alpha^{-1/2}(\Gamma) \oplus H_\alpha^{1/2}(\Gamma)$. Due to the smoothness assumption that Γ is of class $C^{3,\alpha}$, the operators K and $\widetilde{K}$ are compact on $H_\alpha^{1/2}(\Gamma)$ and $H_\alpha^{-1/2}(\Gamma)$, respectively. (Our smoothness assumption is far from optimal at this point, see again [2, Section 3.4].) Hence follows that M_{D} is Fredholm of index zero and M_{D}^* as well [63, Theorem 2.27]. □

At least two features of the factorization

$$N = -LM_{\mathrm{D}}^* L^*$$

are out of the ordinary and need to be treated in a special way: The middle operator M_{D} is indefinite, as S_{D} and T_{D} are coercive for $\omega = \mathrm{i}$. Further, the kernel of L has infinite dimension, as we showed in Proposition II.10. We treat these two problems by a technique similar to the one introduced by Charalambopoulus *et al.* [17] for an inverse transmission problem in elasticity. In contrast to [17], the data operator in our problem is not normal and its non-selfadjoint part is not positive. As in [17] we modify the factorization in such a way that for frequency $\omega = \mathrm{i}$ the middle operator becomes coercive.

In the first step we split up the non-injective part of L, using the Dirichlet-to-Neumann operator $\Upsilon_{(\varepsilon_-,-)}$. This allows to use an injective modification of L. Since the kernel of L contains exactly the pairs $(g, \Upsilon_{(\varepsilon_-,-)} g)$, for $g \in H_\alpha^{1/2}(\Gamma)$, it holds

$$\begin{aligned} L &= \frac{L}{2} \begin{bmatrix} \mathrm{Id} & \Upsilon_{(\varepsilon_-,-)}^{-1} \\ \Upsilon_{(\varepsilon_-,-)} & \mathrm{Id} \end{bmatrix} + \frac{L}{2} \begin{bmatrix} \mathrm{Id} & -\Upsilon_{(\varepsilon_-,-)}^{-1} \\ -\Upsilon_{(\varepsilon_-,-)} & \mathrm{Id} \end{bmatrix} \\ &= \underbrace{\frac{L}{2} \begin{bmatrix} \mathrm{Id} \\ \Upsilon_{(\varepsilon_-,-)} \end{bmatrix}}_{=0} \begin{bmatrix} \mathrm{Id} & \Upsilon_{(\varepsilon_-,-)}^{-1} \end{bmatrix} + \frac{L}{2} \begin{bmatrix} \mathrm{Id} \\ -\Upsilon_{(\varepsilon_-,-)} \end{bmatrix} \begin{bmatrix} \mathrm{Id} & -\Upsilon_{(\varepsilon_-,-)}^{-1} \end{bmatrix} \qquad \text{(II.37)} \\ &= \frac{L}{2} \begin{bmatrix} \mathrm{Id} \\ -\Upsilon_{(\varepsilon_-,-)} \end{bmatrix} \begin{bmatrix} \mathrm{Id} & -\Upsilon_{(\varepsilon_-,-)}^{-1} \end{bmatrix} = \mathfrak{L} \begin{bmatrix} \mathrm{Id} & -\Upsilon_{(\varepsilon_-,-)}^{-1} \end{bmatrix}, \end{aligned}$$

where $\mathfrak{L} : H_\alpha^{1/2}(\Gamma) \to \ell^2$ is defined as $\mathfrak{L} := L/2 \begin{bmatrix} \mathrm{Id} \\ -\Upsilon_{(\varepsilon_-,-)} \end{bmatrix}$. Due to Lemma II.10, $\mathfrak{L}$ is compact and injective and its range is dense in ℓ^2. The factorization of N transforms into

$$N = -\mathfrak{L} \mathcal{M}_{\mathrm{D}}^* \mathfrak{L}^* \qquad \text{with } \mathcal{M}_{\mathrm{D}}^* := \begin{bmatrix} \mathrm{Id} & -\Upsilon_{(\varepsilon_-,-)}^{-1} \end{bmatrix} M_{\mathrm{D}}^* \begin{bmatrix} \mathrm{Id} \\ -\Upsilon_{(\varepsilon_-,-)}^{-1*} \end{bmatrix}. \qquad \text{(II.38)}$$

This defines the bounded operator $\mathcal{M}_{\mathrm{D}} : H_\alpha^{-1/2}(\Gamma) \to H_\alpha^{1/2}(\Gamma)$, the adjoint of which plays the role of the middle operator in the factorization of the near field operator in the remainder of this chapter.

II-6. Properties of the Factorization

The aim of this section is to prepare the construction of a Factorization method based on the factorization of the near field operator. To be able to apply Theorem I.7 on range identities, we are going to show that $\mathcal{M}_{\mathrm{D}}^*$ is a compact perturbation of a coercive operator. Therefore we split up $\mathcal{M}_{\mathrm{D}}$, defined via the Dirichlet Green's function of the half space, in two parts, where the first part consists of layer potential operators defined via the standard quasiperiodic Green's function, and the second part is compact:

$$\begin{aligned}\mathcal{M}_{\mathrm{D}} &= \left[S_{\mathrm{D}}^{\varepsilon_+} + S^{\varepsilon_-}\right] - \left[K_{\mathrm{D}}^{\varepsilon_+} + K^{\varepsilon_-}\right] \Upsilon_{(\varepsilon_-,-)}^{-1*} \\ &\quad - \Upsilon_{(\varepsilon_-,-)}^{-1} \left[\widetilde{K}_{\mathrm{D}}^{\varepsilon_+} + \widetilde{K}^{\varepsilon_-}\right] + \Upsilon_{(\varepsilon_-,-)}^{-1} \left[T_{\mathrm{D}}^{\varepsilon_+} + T^{\varepsilon_-}\right] \Upsilon_{(\varepsilon_-,-)}^{-1*} \\ &= \left[S^{\varepsilon_+} + S^{\varepsilon_+}\right] - \left[K^{\varepsilon_+} + K^{\varepsilon_-}\right] \Upsilon_{(\varepsilon_-,-)}^{-1*} \\ &\quad - \Upsilon_{(\varepsilon_-,-)}^{-1} \left[\widetilde{K}^{\varepsilon_+} + \widetilde{K}^{\varepsilon_-}\right] + \Upsilon_{(\varepsilon_-,-)}^{-1} \left[T^{\varepsilon_+} + T^{\varepsilon_-}\right] \Upsilon_{(\varepsilon_-,-)}^{-1*} + R \\ &= \mathcal{M} + R.\end{aligned} \tag{II.39}$$

The operator R takes all the parts of the layer potential operators S_{D}, K_{D}, $\widetilde{K}_{\mathrm{D}}$ and T_{D} that arise from the reflection at $\{x_2 = 0\}$. Especially, R is an integral operator with smooth kernel and therefore compact from $H_\alpha^{-1/2}(\Gamma)$ into $H_\alpha^{1/2}(\Gamma)$. Let us now denote the dependence of $\mathcal{M}$ on the frequency $\omega \in \mathbb{C}_+$ by $\mathcal{M} = \mathcal{M}_\omega$. In the following, we show that the difference $\mathcal{M}_\omega - \mathcal{M}_{\omega_0}$ is compact and that $\mathcal{M}_\omega$ is coercive for $\omega = \mathrm{i}$.

PROPOSITION II.14. *$\mathcal{M}_\omega - \mathcal{M}_{\omega_0}$ is compact for ω and $\omega_0 \in \mathbb{C}_+ \setminus \mathcal{E}_{\mathrm{d}}$.*

Proof. It is well known that differences of layer operators for different frequencies are compact if the boundary Γ is smooth enough. Our assumption that Γ is of Hölder class $C^{3,\alpha}$ provides enough smoothness, see, e.g., [43, 2]. The compactness of the difference of Dirichlet-to-Neumann operators for different frequencies follows from standard regularity results: Consider for example $\Upsilon = \Upsilon_{(\varepsilon_+,+)}$ and denote the dependence on ω by $\Upsilon = \Upsilon_\omega$. We prove that $\Upsilon_\omega - \Upsilon_{\omega_0}$ is compact from $H_\alpha^{1/2}(\Gamma)$ to $H_\alpha^{-1/2}(\Gamma)$. Let v and v^0 be the upward radiating solutions of the Helmholtz equation with $v|_\Gamma^+ = v^0|_\Gamma^+ = \psi \in H_\alpha^{1/2}(\Gamma)$ and $\Upsilon_\omega \psi = \partial v/\partial\nu$, $\Upsilon_{\omega_0}\psi = \partial v_0/\partial\nu$. The difference $w = v - v^0$ solves

$$\Delta w + \omega^2 \varepsilon_+ w = \varepsilon_+(\omega_0^2 - \omega^2) v_0 \quad \text{in } \Omega_+, \qquad w = 0 \quad \text{on } \Omega_-.$$

Interior regularity results [63, Theorem 4.16] imply that w belongs to $H_\alpha^2\big((0,2\pi)\times(h,3h)\big)$, with continuous dependence on the norm of ψ,

$$\|w\|_{H_\alpha^2\left((0,2\pi)\times(h,3h)\right)} \le C\|w\|_{H_\alpha^1\left(D_+\cup(0,2\pi)\times[h,3h)\right)} + C\|v_0\|_{L^2\left(D_+\cup(0,2\pi)\times[h,3h)\right)} \le C\|\psi\|_{H_\alpha^{1/2}(\Gamma)}.$$

Especially, the boundary values $w|_{x_2=2h}$ belong to $H_\alpha^{3/2}(\{x_2 = 2h\})$. Thus, global regularity results [63, Theorem 4.18] yield that $w \in H_\alpha^2(D_+)$,

$$\|w\|_{H_\alpha^2(D_+)} \le C\|w\|_{H_\alpha^1(D_+)} + C\|w\|_{H_\alpha^{3/2}(\{x_2=2h\})} + C\|v_0\|_{L^2(D_+)},$$

and therefore $\partial(v - v^0)/\partial\nu|_\Gamma^+ \in H_\alpha^{1/2}(\Gamma)$. The compact embedding of $H_\alpha^{1/2}(\Gamma)$ in $H_\alpha^{-1/2}(\Gamma)$ implies compactness of $\Upsilon_\omega - \Upsilon_{\omega_0}$. The Neumann-to-Dirichlet operator $\Upsilon^{-1}_{(\varepsilon_+,+)}$ can be treated be similar arguments. □

Proposition II.14 is the only part of this work where we need the strong smoothness assumptions on Γ. All other arguments are valid even for a Lipschitz surface.

In the remainder of this section, we show that $\mathcal{M}$ (or equivalently $\mathcal{M}_\mathrm{D}$) is a compact perturbation of a coercive operator. To prove this result, we first split $\mathcal{M}$ in two parts, $\mathcal{M} = \mathcal{M}^+ + \mathcal{M}^-$,

$$\mathcal{M}^\pm = S^{\varepsilon_\pm} - K^{\varepsilon_\pm}\Upsilon^{-1*}_{(\varepsilon_-,-)} - \Upsilon^{-1}_{(\varepsilon_-,-)}\widetilde{K}^{\varepsilon_\pm} - \Upsilon^{-1}_{(\varepsilon_-,-)} T^{\varepsilon_\pm}\Upsilon^{-1*}_{(\varepsilon_-,-)}.$$

Next we show in a sequel of lemmata that $\mathcal{M}^+ = \mathcal{M}_\mathrm{i}^+$ is coercive for $\omega = \mathrm{i}$. Since the differences $\mathcal{M}^- - \mathcal{M}_\mathrm{i}^-$, $\mathcal{M}_\mathrm{i}^- - \mathcal{M}_\mathrm{i}^+$ and $\mathcal{M}^+ - \mathcal{M}_\mathrm{i}^+$ are compact it follows that

$$\mathcal{M} = \mathcal{M}^+ + \mathcal{M}^- = (\mathcal{M}^+ - \mathcal{M}_\mathrm{i}^+) + (\mathcal{M}^- - \mathcal{M}_\mathrm{i}^-) + (\mathcal{M}_\mathrm{i}^- - \mathcal{M}_\mathrm{i}^+) + 2\mathcal{M}_\mathrm{i}^+ \quad \text{(II.40)}$$

is a compact perturbation of a coercive operator. Hence, we restrict us in the following to the case $\omega = \mathrm{i}$ and denote by $\mathrm{SL}_\mathrm{i}^{\varepsilon_\pm}$ and $S_\mathrm{i}^{\varepsilon_\pm}$ the single layer potential and operator on Γ for the wavenumber $\omega^2\varepsilon_\pm = -\varepsilon_\pm$, respectively.

Lemma II.15 (Lemmas 3.3, 3.4 in [17]). *For* $\omega = \mathrm{i}$ *the operator* $B := \Upsilon_{(\varepsilon_-,-)} - \Upsilon_{(\varepsilon_+,+)} : H^{1/2}_\alpha(\Gamma) \to H^{-1/2}_\alpha(\Gamma)$ *is coercive and selfadjoint and*

$$\Upsilon_{(\varepsilon_-,-)} \mathcal{M}^+_{\mathrm{i}} \Upsilon_{(\varepsilon_-,-)} = B\, S^{\varepsilon+}_{\mathrm{i}}\, B - B. \tag{II.41}$$

Note that for $\omega = \mathrm{i}$, $\Upsilon_{(\varepsilon_\pm,\pm)}$ and $\Lambda^\pm$ are selfadjoint and coercive operators. This follows from the representation (II.7) and the second part of the proof of Lemma II.9. The single layer operator $S^{\varepsilon+}_{\mathrm{i}} H^{-1/2}_\alpha(\Gamma) \to H^{1/2}_\alpha(\Gamma)$ is also selfadjoint for frequency i, since its kernel is real and symmetric. Hence we observe from (II.41) that $\mathcal{M}^+_{\mathrm{i}}$ is selfadjoint for $\omega = \mathrm{i}$, too.

Proof. The proof of the decomposition follows as in the proof of [17, Lemma 3.3] using Calderón identities. For $\omega = \mathrm{i}$ the single layer operator $S^{\varepsilon\pm}_{\mathrm{i}}$ is selfadjoint and coercive (coercivity has already been shown in the proof of Theorem II.13). From the jump relations for the single layer potential in Theorem II.6 we conclude the identity

$$\Upsilon_{(\varepsilon_+,+)}\psi - \Upsilon_{(\varepsilon_+,-)}\psi = \left[\frac{\partial}{\partial\nu}\, \mathrm{SL}^{\varepsilon+}_{\mathrm{i}}\, (S^{\varepsilon+}_{\mathrm{i}})^{-1}\psi\right]_\Gamma = -(S^{\varepsilon+}_{\mathrm{i}})^{-1}\psi,$$

for $\psi \in H^{1/2}_\alpha(\Gamma)$, or equivalently,

$$S^{\varepsilon+}_{\mathrm{i}} \left(\Upsilon_{(\varepsilon_+,-)} - \Upsilon_{(\varepsilon_+,+)}\right) = \mathrm{Id}\,. \tag{II.42}$$

Moreover, the jump relations also imply that $\Upsilon_{(\varepsilon_+,-)} S^{\varepsilon+}_{\mathrm{i}} = \mathrm{Id}/2 + \widetilde{K}^{\varepsilon+}_{\mathrm{i}}$ and due to selfadjointness of $\Upsilon_{(\varepsilon_+,-)}$ and $S^{\varepsilon+}_{\mathrm{i}}$ also $S^{\varepsilon+}_{\mathrm{i}} \Upsilon_{(\varepsilon_+,-)} = \mathrm{Id}/2 + K^{\varepsilon+}_{\mathrm{i}}$. Finally, the Calderón identities $\widetilde{K}^{\varepsilon+}_{\mathrm{i}} \Upsilon_{(\varepsilon_+,-)} = \Upsilon_{(\varepsilon_+,-)} K^{\varepsilon+}_{\mathrm{i}}$ and $\Upsilon_{(\varepsilon_+,-)} = T(\mathrm{Id}/2 - K^{\varepsilon+}_{\mathrm{i}})$ imply with the help of the same computation as it is performed in the proof of [17, Lemma 3.3] the equality claimed in (II.41).

Now we show coercivity of B for $\omega = \mathrm{i}$, following the ideas of [17, Lemma 3.4]. The arguments of Lemma 3.4 in [17] are not directly applicable here due to the different character of Kupradze's radiation condition and the Rayleigh condition (II.3). For $\psi \in H^{1/2}_\alpha(\Gamma)$ we set $u = \mathrm{SL}^{\varepsilon+}_{\mathrm{i}} (\mathrm{S}^{\varepsilon+}_{\mathrm{i}})^{-1}\psi \in$

$H^1_{\alpha,\mathrm{loc}}(\Pi)$ and $v = \mathrm{SL}_{\mathrm{i}}^{\varepsilon_-}(S_{\mathrm{i}}^{\varepsilon_-})^{-1}\psi \in H^1_{\alpha,\mathrm{loc}}(\Pi)$. Then

$$\begin{aligned}
\int_\Gamma \overline{\psi} B\psi \,\mathrm{d}s &= \int_\Gamma \overline{\psi}\Upsilon_{(\varepsilon_-,-)}\psi \,\mathrm{d}s - \int_\Gamma \overline{\psi}\Upsilon_{(\varepsilon_+,+)}\psi \,\mathrm{d}s \\
&= \int_\Gamma \overline{v}|_\Gamma^- \left.\frac{\partial v}{\partial \nu}\right|_\Gamma^- \mathrm{d}s - \int_\Gamma \overline{u}|_\Gamma^+ \left.\frac{\partial u}{\partial \nu}\right|_\Gamma^+ \mathrm{d}s \\
&= \int_{D_-} |\nabla v|^2 + \varepsilon_- |v|^2 \,\mathrm{d}x - \int_{\Gamma^-} \overline{v}\Lambda^- v \,\mathrm{d}s \\
&\quad + \int_{D_+} |\nabla u|^2 + \varepsilon_+ |u|^2 \,\mathrm{d}x - \int_{\Gamma^+} \overline{u}\Lambda^+ u \,\mathrm{d}s \\
&\geq C\|u\|^2_{H^1_\alpha(D_+)} + C\|v\|^2_{H^1_\alpha(D_-)} \geq C\|\psi\|^2_{H^{1/2}_\alpha(\Gamma)}.
\end{aligned}$$

□

Coercivity of B for the frequency $\omega = \mathrm{i}$ yields positivity of $\mathcal{M}_{\mathrm{i}}^+$, as the next proposition shows. The proof is a modification of [17, Lemma 3.5]. The main difference compared to [17] is here that the coefficients $\varepsilon_\pm$ appear merely in the lower order terms.

PROPOSITION II.16. *For $\omega = \mathrm{i}$ the operator $\mathcal{M}_{\mathrm{i}}^+$ is positive for $0 < \varepsilon_+ < \varepsilon_-$ and negative for $0 < \varepsilon_- < \varepsilon_+$.*

Proof. For $\omega = \mathrm{i}$, $\Upsilon_{(\varepsilon_-,-)}$ is selfadjoint and Lemma II.15 implies that it is sufficient to show that $A := BS_{\mathrm{i}}^{\varepsilon_+}B - B$ is positive on $H^{1/2}_\alpha(\Gamma)$. For $\psi \in H^{1/2}_\alpha(\Gamma)$ we define

$$v := \mathrm{SL}_{\mathrm{i}}^{\varepsilon_+}(B\psi), \quad w_+ := \mathrm{SL}_{\mathrm{i}}^{\varepsilon_+}(S_{\mathrm{i}}^{\varepsilon_+})^{-1}\psi, \quad w_- := \mathrm{SL}_{\mathrm{i}}^{\varepsilon_-}(S_{\mathrm{i}}^{\varepsilon_-})^{-1}\psi,$$

and note that $\psi = w_+|_\Gamma^\pm = w_-|_\Gamma^\pm$. Moreover,

$$B\psi = -\left[\frac{\partial v}{\partial \nu}\right]_\Gamma, \quad B\psi = \Upsilon_{(\varepsilon_-,-)}\psi - \Upsilon_{(\varepsilon_+,+)}\psi = \left.\frac{\partial w_-}{\partial \nu}\right|_\Gamma^- - \left.\frac{\partial w_+}{\partial \nu}\right|_\Gamma^+.$$

These two representations of $B\psi$ yield four representations of

$$\int_\Gamma \overline{\psi} A\psi \,\mathrm{d}s = \int_\Gamma \overline{B\psi}\, S^{\varepsilon_+} B\psi \,\mathrm{d}s - \int_\Gamma \overline{\psi}\, B\psi \,\mathrm{d}s$$

that we write up, for simplicity, using the sesquilinear form

$$\Psi_\varepsilon^\pm(v,w) := \int_{D_\pm} \nabla v \nabla \overline{w} + \varepsilon v \overline{w}\,\mathrm{d}x - \int_{\Gamma_\pm} \overline{w}\Lambda^\pm v\,\mathrm{d}s, \qquad \varepsilon \in \{\varepsilon_+, \varepsilon_-\},$$

for $v, w \in H^1_\alpha(D_\pm)$. Note that $\Psi_\varepsilon^\pm$ is antisymmetric in v and w. As in [17, Proof of Lemma 3.5] we find

$$\begin{aligned}\int_\Gamma \overline{\psi} A\psi\,\mathrm{d}s &= \int_\Gamma \overline{B\psi} S^{\varepsilon_+} B\psi\,\mathrm{d}s - \int_\Gamma \overline{\psi} B\psi\,\mathrm{d}s \\ &= \int_\Gamma v \left[\frac{\partial w_-}{\partial\nu}\Big|_\Gamma^- - \frac{\partial w_+}{\partial\nu}\Big|_\Gamma^+ \right] \mathrm{d}s - \int_\Gamma \left[\frac{\partial v}{\partial\nu}\Big|_\Gamma^- - \frac{\partial v}{\partial\nu}\Big|_\Gamma^+ \right] \overline{w}_+\,\mathrm{d}s.\end{aligned}$$

Now, since $w_+|_\Gamma = w_-|_\Gamma$, Green's first identity implies

$$\begin{aligned}\int_\Gamma \overline{\psi} A\psi\,\mathrm{d}s &= \Psi^-_{\varepsilon_-}(v,w_-) + \Psi^+_{\varepsilon_+}(v,w_+) - \Psi^-_{\varepsilon_+}(v,w_-) - \Psi^+_{\varepsilon_+}(v,w_+) \\ &= \Psi^-_{\varepsilon_-}(v,w_-) - \Psi^-_{\varepsilon_+}(v,w_-). \end{aligned} \tag{II.43}$$

Exploiting the two different representations of $B\psi$, we also obtain

$$\begin{aligned}\int_\Gamma \overline{\psi} A\psi\,\mathrm{d}s &= \int_\Gamma v \left[\frac{\partial v}{\partial\nu}\Big|_\Gamma^- - \frac{\partial v}{\partial\nu}\Big|_\Gamma^+ \right] \mathrm{d}s - \int_\Gamma \left[\frac{\partial v}{\partial\nu}\Big|_\Gamma^- - \frac{\partial v}{\partial\nu}\Big|_\Gamma^+ \right] \overline{w}_+\,\mathrm{d}s \\ &= \Psi^-_{\varepsilon_+}(v,v) + \Psi^+_{\varepsilon_+}(v,v) - \Psi^-_{\varepsilon_+}(v,w_-) - \Psi^+_{\varepsilon_+}(v,w_+) \\ &= \Psi^-_{\varepsilon_+}(v,v-w_-) + \Psi^+_{\varepsilon_+}(v,v-w_+), \end{aligned} \tag{II.44}$$

and finally

$$\begin{aligned}\int_\Gamma \overline{\psi} A\psi\,\mathrm{d}s &= \int_\Gamma v \left[\frac{\partial v}{\partial\nu}\Big|_\Gamma^- - \frac{\partial v}{\partial\nu}\Big|_\Gamma^+ \right] \mathrm{d}s - \int_\Gamma \left[\frac{\partial w_-}{\partial\nu}\Big|_\Gamma^- - \frac{\partial w_+}{\partial\nu}\Big|_\Gamma^+ \right] \overline{w}_+\,\mathrm{d}s \\ &= \Psi^-_{\varepsilon_-}(v,w_-) + \Psi^+_{\varepsilon_+}(v,w_+) - \Psi^-_{\varepsilon_-}(w_-,w_-) - \Psi^+_{\varepsilon_+}(w_+,w_+) \\ &= \Psi^-_{\varepsilon_-}(v-w_-,w_-) + \Psi^+_{\varepsilon_+}(v-w_+,w_+). \end{aligned} \tag{II.45}$$

Subtracting (II.45) from (II.44) yields

$$0 = \Psi^-_{\varepsilon_+}(v,v-w_-) - \Psi^-_{\varepsilon_-}(w_-,v-w_-) + \Psi^+_{\varepsilon_+}(v-w_+,v-w_+) \tag{II.46}$$

and thus

$$0 = \Psi^-_{\varepsilon_-}(v - w_-, v - w_-) + \Psi^+_{\varepsilon_+}(v - w_+, v - w_+) \\ - \Psi^-_{\varepsilon_-}(v, v - w_-) + \Psi^-_{\varepsilon_+}(v, v - w_-). \quad \text{(II.47)}$$

Subtracting (II.43) from (II.47) results in

$$- \int_\Gamma \overline{\psi} A \psi \, \mathrm{d}s = \Psi^-_{\varepsilon_+}(v, v) - \Psi^-_{\varepsilon_-}(v, v) \\ + \Psi^+_{\varepsilon_+}(v - w_+, v - w_+) + \Psi^-_{\varepsilon_-}(v - w_-, v - w_-). \quad \text{(II.48)}$$

Moreover, adding (II.43) to (II.46) yields

$$\begin{aligned} \int_\Gamma \overline{\psi} A \psi \, \mathrm{d}s &= \Psi^-_{\varepsilon_-}(v, w_-) - \Psi^-_{\varepsilon_+}(v, w_-) + \Psi^-_{\varepsilon_+}(v, v - w_-) \\ &\quad - \Psi^-_{\varepsilon_-}(w_-, v - w_-) + \Psi^+_{\varepsilon_+}(v - w_+, v - w_+) \\ &= \Psi^-_{\varepsilon_-}(w_-, w_-) - \Psi^-_{\varepsilon_+}(v, w_-) + \Psi^-_{\varepsilon_+}(v - w_-, v - w_-) \\ &\quad + \Psi^-_{\varepsilon_+}(w_-, v - w_-) + \Psi^+_{\varepsilon_+}(v - w_+, v - w_+) \\ &\quad + 2\mathrm{i}\,\mathrm{Im}\left(\Psi^-_{\varepsilon_-}(v, w_-)\right) \\ &= \Psi^-_{\varepsilon_-}(w_-, w_-) - \Psi^-_{\varepsilon+}(w_-, w_-) \\ &\quad + \Psi^-_{\varepsilon_+}(v - w_-, v - w_-) + \Psi^+_{\varepsilon_+}(v - w_+, v - w_+) \\ &\quad + 2\mathrm{i}\,\mathrm{Im}\left(\Psi^-_{\varepsilon_-}(v, w_-)\right) + 2\mathrm{i}\,\mathrm{Im}\left(\Psi^-_{\varepsilon_+}(w_-, v)\right). \end{aligned}$$

Taking the real part of the last equation, the selfadjointness of A implies

$$\int_\Gamma \overline{\psi} A \psi \, \mathrm{d}s = \Psi^-_{\varepsilon_-}(w_-, w_-) - \Psi^-_{\varepsilon+}(w_-, w_-) \\ + \Psi^-_{\varepsilon_+}(v - w_-, v - w_-) + \Psi^+_{\varepsilon_+}(v - w_+, v - w_+). \quad \text{(II.49)}$$

The difference $\Psi^-_{\varepsilon_+} - \Psi^-_{\varepsilon_-}$ is positive for $0 < \varepsilon_- < \varepsilon_+$, since

$$\Psi^-_{\varepsilon_+}(v, v) - \Psi^-_{\varepsilon_-}(v, v) = (\varepsilon_+ - \varepsilon_-) \int_{D_-} |v|^2 \, \mathrm{d}x \\ + 2\pi \sum_{n \in \mathbb{Z}} \left(\sqrt{\varepsilon_+ + \alpha_n^2} - \sqrt{\varepsilon_- + \alpha_n^2} \right) |\hat{v}_n|^2 > 0$$

for $v \neq 0$. Here, $\hat{v}_n$ denote the nth Fourier coefficient of $\exp(-\mathrm{i}\alpha\cdot)v|_{\Gamma_-}$. Hence, equation (II.48) shows that $\mathcal{M}_\mathrm{i}^+$ is negative definite. On the other hand, for $0 < \varepsilon_+ < \varepsilon_-$ one observes by a similar computation that $\Psi_{\varepsilon_-}^-(w_-, w_-) - \Psi_{\varepsilon_+}^-(w_-, w_-)$ is positive and (II.49) implies that $\mathcal{M}_\mathrm{i}^+$ is positive definite. This was the claim of the proposition. □

PROPOSITION II.17. *For* $\omega = \mathrm{i}$ *and* $0 < \varepsilon_+ < \varepsilon_-$ *the operator* $\mathcal{M}_\mathrm{i}^+ : H_\alpha^{-1/2}(\Gamma) \to H_\alpha^{1/2}(\Gamma)$ *is positive definite. In case that* $0 < \varepsilon_- < \varepsilon_+$, $-\mathcal{M}_\mathrm{i}^+$ *is positive definite.*

Proof. We only consider the case $0 < \varepsilon_+ < \varepsilon_-$ in the proof, such that $\mathcal{M}_\mathrm{i}^+$ is positive definite. The case $0 < \varepsilon_- < \varepsilon_+$ can be treated analogously.

Since $Q := [\mathrm{Id}, -\Upsilon_{(\varepsilon_-,-)}^{-1*}]^\top$ is an isomorphism, the Fredholm index theorem [63, Theorem 2.21] yields that

$$\mathcal{M}_\mathrm{i}^+ = Q^* \begin{pmatrix} S_\mathrm{i}^{\varepsilon_+} & K_\mathrm{i}^{\varepsilon_+} \\ \widetilde{K}_\mathrm{i}^{\varepsilon_+} & -T_\mathrm{i}^{\varepsilon_+} \end{pmatrix} Q$$

is a Fredholm operator of index 0, compare Proposition II.13. Moreover, $\mathcal{M}_\mathrm{i}^+$ is a positive and hence an injective operator due to the last Proposition II.16. Especially, $\mathcal{M}_\mathrm{i}^+$ is an isomorphism. Since $\mathcal{M}_\mathrm{i}^+$ is also selfadjoint (see the remark directly after Lemma II.15), we can exploit (I.28) in the same way as in Chapter I, yielding

$$\int_\Gamma \overline{\phi} \mathcal{M}_\mathrm{i}^+ \phi \,\mathrm{d}s \geq \frac{1}{\|\mathcal{M}_\mathrm{i}^+\|} \|\mathcal{M}_\mathrm{i}^+ \phi\|^2 \geq \frac{C}{\|\mathcal{M}_\mathrm{i}^+\|} \|\phi\|^2, \qquad \phi \in H_\alpha^{-1/2}(\Gamma).$$

Therefore we obtain that $\mathcal{M}_\mathrm{i}^+$ is a positive definite operator. □

COROLLARY II.18. *For* $\varepsilon_+ \neq \varepsilon_-$, *the real part* $\mathrm{Re}\, \mathcal{M}_\mathrm{D}^*$ *of the middle operator* $\mathcal{M}_\mathrm{D}^*$ *of the factorization* (II.38) *is a compact perturbation of a coercive operator.*

Proof. The splittings (II.39) and (II.40) show that $\mathrm{Re}\, \mathcal{M}_\mathrm{D}^* = 1/2(\mathcal{M}_\mathrm{D} + \mathcal{M}_\mathrm{D}^*)$ is a compact perturbation of $2\mathcal{M}_\mathrm{i}^+$. Depending on the sign of $\varepsilon_+ - \varepsilon_-$, the latter operator has been shown to be positive or negative definite in the last proposition, thus, $\mathcal{M}_\mathrm{i}^+$ is coercive in case that $\varepsilon_+ \neq \varepsilon_-$. □

II-7. A Factorization Method

In Corollary II.18 we showed that $\operatorname{Re}\mathcal{M}_{\mathrm{D}}^*$ is a compact perturbation of a coercive operator. This result is the basis for the Factorization method which we construct now. As we know from the first chapter, at least semidefiniteness of the middle operator $\operatorname{Im}\mathcal{M}_{\mathrm{D}}^*$ is necessary to apply Theorem I.7 to the factorization $N = -\mathcal{L}\mathcal{M}_{\mathrm{D}}^*\mathcal{L}^*$. In the first part of this section we show that the middle operator is indeed semidefinite, but not definite. Therefore our version of the basic result on range identities in Theorem I.7 can be applied, but former versions of this theorem do not apply such easily, as we explain afterwards.

The middle operator $\operatorname{Im}\mathcal{M}_{\mathrm{D}}^*$ of the factorization of N is negative semidefinite if and only if $\operatorname{Im}\langle \mathcal{M}_{\mathrm{D}}^*\phi, \phi\rangle \le 0$ for all $\phi \in H_\alpha^{-1/2}(\Gamma)$. Since the range of the operator $\mathcal{L}^*$ from (II.37) is dense in $H_\alpha^{1/2}(\Gamma)$, it is sufficient to show that $\operatorname{Im}\langle \mathcal{L}\mathcal{M}_{\mathrm{D}}^*\mathcal{L}^*(a_n), (a_n)\rangle \le 0$ for all $(a_n) \in \ell^2$. Moreover, $\mathcal{L}\mathcal{M}_{\mathrm{D}}^*\mathcal{L}^* = LM_{\mathrm{D}}^*L^*$ and (compare the discussion after (II.35))

$$LM_{\mathrm{D}}^*L^* = L\begin{pmatrix} S_{\mathrm{D}}^{\varepsilon+} & K_{\mathrm{D}}^{\varepsilon+} \\ \widetilde{K}_{\mathrm{D}}^{\varepsilon+} & -T_{\mathrm{D}}^{\varepsilon+}\end{pmatrix}^* L^*.$$

Therefore we show in the following that

$$\begin{aligned}\operatorname{Im}&\left\langle \begin{pmatrix} S_{\mathrm{D}}^{\varepsilon+} & K_{\mathrm{D}}^{\varepsilon+} \\ \widetilde{K}_{\mathrm{D}}^{\varepsilon+} & -T_{\mathrm{D}}^{\varepsilon+}\end{pmatrix}\begin{pmatrix}\phi\\ \psi\end{pmatrix}, \begin{pmatrix}\phi\\ \psi\end{pmatrix}\right\rangle \\ &= \operatorname{Im}\Big(\langle S_{\mathrm{D}}^{\varepsilon+}\phi,\phi\rangle + \langle K_{\mathrm{D}}^{\varepsilon+}\psi,\phi\rangle + \langle \widetilde{K}_{\mathrm{D}}^{\varepsilon+}\phi,\psi\rangle - \langle T_{\mathrm{D}}^{\varepsilon+}\psi,\psi\rangle\Big) \ge 0\end{aligned}$$

for $\phi \in H_\alpha^{-1/2}(\Gamma)$ and $\psi \in H_\alpha^{1/2}(\Gamma)$. For this computation we find it convenient to denote the inner product on $L^2(\Gamma)$ by $\langle\cdot,\cdot\rangle$. Let $u = \mathrm{SL}_{\mathrm{D}}^{\varepsilon+}\phi$ and $v = \mathrm{DL}_{\mathrm{D}}^{\varepsilon+}\psi$ for $\phi \in H_\alpha^{-1/2}(\Gamma)$ and $\psi \in H_\alpha^{1/2}(\Gamma)$. Then Green's first identity shows that

$$\begin{aligned}\langle S_{\mathrm{D}}^{\varepsilon+}\phi,\phi\rangle &= \left\langle u, -\left[\frac{\partial u}{\partial\nu}\right]_\Gamma\right\rangle \\ &= \int_{D_+\cup D_-}\left(|\nabla u|^2 - k^2|u|^2\right)\mathrm{d}x - \int_{\Gamma_+} u\overline{\Lambda_{\varepsilon_+}^+ u}\,\mathrm{d}s\end{aligned}$$

and

$$\langle T_{\mathrm{D}}^{\varepsilon+}\psi,\psi\rangle = \left\langle -\frac{\partial v}{\partial \nu}, [v]_\Gamma \right\rangle$$
$$= \int_{D_+\cup D_-} \left(|\nabla v|^2 - k^2|v|^2\right) \mathrm{d}x - \int_{\Gamma_+} \overline{v}\Lambda_{\varepsilon_+}^{+} v \,\mathrm{d}s.$$

We conclude that

$$\mathrm{Im}\,\langle S_{\mathrm{D}}^{\varepsilon+}\phi,\phi\rangle = -\mathrm{Im}\int_{\Gamma_+} u\overline{\Lambda_{\varepsilon_+}^{+} u}\,\mathrm{d}s = \mathrm{Im}\int_{\Gamma_+} \overline{u}\Lambda_{\varepsilon_+}^{+} u\,\mathrm{d}s \geq 0$$

and $-\mathrm{Im}\,\langle T_{\mathrm{D}}^{\varepsilon+}\psi,\psi\rangle = \mathrm{Im}\int_{\Gamma_+}\overline{v}\Lambda_{\varepsilon_+}^{+} v\,\mathrm{d}s \geq 0$. The term $\mathrm{Im}\int_{\Gamma_+}\overline{v}\Lambda_{\varepsilon_+}^{+} v\,\mathrm{d}s$ is non-negative due to the representation of $\Lambda_{\varepsilon_+}^{+}$ in (II.7). We furthermore compute

$$\langle K_{\mathrm{D}}^{\varepsilon+}\psi,\phi\rangle + \langle \widetilde{K}_{\mathrm{D}}^{\varepsilon+}\phi,\psi\rangle = \left\langle \mathrm{DL}_{\mathrm{D}}^{\varepsilon+}\psi\big|_\Gamma^+ + \mathrm{DL}_{\mathrm{D}}^{\varepsilon+}\psi\big|_\Gamma^-, -\left[\frac{\partial}{\partial\nu}\mathrm{SL}_{\mathrm{D}}^{\varepsilon+}\phi\right]_\Gamma\right\rangle$$
$$+\left\langle \frac{\partial}{\partial\nu}\mathrm{SL}_{\mathrm{D}}^{\varepsilon+}\phi\Big|_\Gamma^+ + \frac{\partial}{\partial\nu}\mathrm{SL}_{\mathrm{D}}^{\varepsilon+}\phi\Big|_\Gamma^-, [\mathrm{DL}_{\mathrm{D}}^{\varepsilon+}\psi]_\Gamma\right\rangle$$
$$= \left\langle v|_\Gamma^+ + v|_\Gamma^-, -\left[\frac{\partial u}{\partial\nu}\right]_\Gamma\right\rangle + \left\langle \frac{\partial u}{\partial\nu}\Big|_\Gamma^+ + \frac{\partial u}{\partial\nu}\Big|_\Gamma^-, v|_\Gamma^+ - v|_\Gamma^-\right\rangle.$$

In the next step, we take the imaginary part of the latter expresssion. Since we use the Dirichlet potentials $\mathrm{SL}_{\mathrm{D}}^{\varepsilon+}$ and $\mathrm{DL}_{\mathrm{D}}^{\varepsilon+}$, both functions u and v solve the Helmholtz equation in D_- and satisfy a homogeneous Dirichlet boundary condition on the interval $\{(x_1,x_2)\in\Pi,\, x_2=0\}$. Therefore Green's second identity shows that $\langle v|_\Gamma^-, \partial u/\partial\nu|_\Gamma^-\rangle - \langle \partial u/\partial\nu|_\Gamma^-, v|_\Gamma^-\rangle$ vanishes. Moreover, the two terms $\langle v|_\Gamma^-, -\partial u/\partial\nu|_\Gamma^+\rangle + \langle \partial u/\partial\nu|_\Gamma^+, -v|_\Gamma^-\rangle$ and $\langle v|_\Gamma^+, \partial u/\partial\nu|_\Gamma^-\rangle + \langle \partial u/\partial\nu|_\Gamma^+, v|_\Gamma^+\rangle$ are real numbers and hence negligible when we consider the imaginary part of $\langle K_{\mathrm{D}}^{\varepsilon+}\psi,\phi\rangle + \langle\widetilde{K}_{\mathrm{D}}^{\varepsilon+}\phi,\psi\rangle$. Thus, by Green's first identity we find

$$\mathrm{Im}\,\langle K_{\mathrm{D}}^{\varepsilon+}\psi,\phi\rangle + \mathrm{Im}\,\langle\widetilde{K}_{\mathrm{D}}^{\varepsilon+}\phi,\psi\rangle = -\mathrm{Im}\left\langle v|_\Gamma^+, \frac{\partial u}{\partial\nu}\Big|_\Gamma^+\right\rangle + \mathrm{Im}\left\langle \frac{\partial u}{\partial\nu}\Big|_\Gamma^+, v|_\Gamma^+\right\rangle$$
$$= -\mathrm{Im}\int_{\Gamma_A} v\overline{\frac{\partial u}{\partial x_2}}\,\mathrm{d}s + \mathrm{Im}\int_{\Gamma_A}\overline{v}\frac{\partial u}{\partial x_2}\,\mathrm{d}s = 2\,\mathrm{Im}\int_{\Gamma_A}\overline{v}\frac{\partial u}{\partial x_2}\,\mathrm{d}s.$$

Here, $\Gamma_A = \{x \in \Pi,\, x_2 = A\}$ and $A > h$ is arbitrary. The Rayleigh expansion of

$$u(x) = \sum_{n\in\mathbb{Z}} u_n^+ \exp(\mathrm{i}(\alpha_n x_1 + \beta_n^+ x_2)), \qquad x_2 > h,$$

and

$$v(x) = \sum_{m\in\mathbb{Z}} v_m^+ \exp(\mathrm{i}(\alpha_m x_1 + \beta_m^+ x_2)), \qquad x_2 > h,$$

reveals that

$$\int_{\Gamma_A} \overline{v} \frac{\partial u}{\partial x_2}\, \mathrm{d}s = \sum_{n,m\in\mathbb{Z}} \mathrm{i}\beta_n^+ u_n^+ \overline{v_m^+} \int_{\Gamma_A} \overline{e^{\mathrm{i}(\alpha_m x_1 + \beta_m^+ x_2)}} e^{\mathrm{i}(\alpha_n x_1 + \beta_n^+ x_2)}\, \mathrm{d}s$$

and hence

$$\begin{aligned} \operatorname{Im} \int_{\Gamma_A} \overline{v} \frac{\partial u}{\partial x_2}\, \mathrm{d}s &= \operatorname{Im} \left[\sum_{n,m\in\mathbb{Z}} \mathrm{i}\beta_n^+ u_n^+ \overline{v_m^+}\, \overline{e^{\mathrm{i}\beta_m^+ A}} e^{\mathrm{i}\beta_n^+ A} \int_{\Gamma_A} e^{\mathrm{i}(n-m)x_1}\, \mathrm{d}s \right] \\ &= 2\pi \operatorname{Im} \left[\sum_{n\in\mathbb{Z}} \mathrm{i}\beta_n^+ u_n^+ \overline{v_n^+}\, \overline{e^{\mathrm{i}\beta_n^+ A}} e^{\mathrm{i}\beta_n^+ A} \right]. \end{aligned} \qquad \text{(II.50)}$$

Note that

$$\begin{aligned} \overline{e^{\mathrm{i}\beta_n^+ A}} e^{\mathrm{i}\beta_n^+ A} &= \overline{e^{\mathrm{i}\sqrt{\omega^2\varepsilon_+ - \alpha_n^2} A}} e^{\mathrm{i}\sqrt{\omega^2\varepsilon_+ - \alpha_n^2} A} \\ &= \begin{cases} 1, & \omega^2\varepsilon_+ \geq \alpha_n^2, \\ e^{2\mathrm{i}\beta_n^+ A}, & \omega^2\varepsilon_+ < \alpha_n^2. \end{cases} \end{aligned}$$

The condition $\omega^2\varepsilon_+ \geq \alpha_n^2$ is satisfied precisely if β_n^+ is a real number which shows that the corresponding mode in the Rayleigh expansion is a propagating one. Otherwise, if $\omega^2\varepsilon_+ < \alpha_n^2$, the corresponding mode in the Rayleigh expansion is evanescent. Therefore, taking the limit as $A \to \infty$ in (II.50) it follows that

$$\begin{aligned} \operatorname{Im} \langle K_{\mathrm{D}}^{\varepsilon_+} \psi, \phi \rangle + \operatorname{Im} \langle \widetilde{K}_{\mathrm{D}}^{\varepsilon_+} \phi, \psi \rangle &= 2 \operatorname{Im} \int_{\Gamma_A} \overline{v} \frac{\partial u}{\partial x_2}\, \mathrm{d}s \\ &= 4\pi \operatorname{Im} \left[\sum_{n\in\mathbb{Z}:\, \beta_n\in\mathbb{R}} \mathrm{i}\beta_n^+ u_n^+ \overline{v_n^+} \right]. \end{aligned}$$

Collecting terms, we arrive at

$$\begin{aligned}
&\langle S_{\mathrm{D}}^{\varepsilon_+}\phi,\phi\rangle + \langle K_{\mathrm{D}}^{\varepsilon_+}\psi,\phi\rangle + \langle \widetilde{K}_{\mathrm{D}}^{\varepsilon_+}\phi,\psi\rangle - \langle T_{\mathrm{D}}^{\varepsilon_+}\psi,\psi\rangle \\
&\quad = \operatorname{Im}\int_{\Gamma_+}\left(\overline{u}\Lambda_{\varepsilon_+}^+ u + \overline{v}\Lambda_{\varepsilon_+}^+ v\right)\,\mathrm{d}s + 4\pi\operatorname{Im}\left[\sum_{n\in\mathbb{Z}\colon\beta_n\in\mathbb{R}} \mathrm{i}\beta_n^+ u_n^+\overline{v_n^+}\right] \\
&\quad = \operatorname{Im}\left(2\pi\mathrm{i}\sum_{j\in\mathbb{Z}}\beta_n^+\left[|u_n^+|^2+|v_n^+|^2\right]\right) + 4\pi\sum_{n\in\mathbb{Z}\colon\beta_n\in\mathbb{R}}\beta_n^+\operatorname{Re}\left[u_n^+\overline{v_n^+}\right] \\
&\quad = 2\pi\sum_{j\in\mathbb{Z}\colon\beta_n\in\mathbb{R}}\beta_n^+\left[|u_n^+|^2 + 2\operatorname{Re}\left(u_n^+\overline{v_n^+}\right) + |v_n^+|^2\right] \\
&\quad = 2\pi\sum_{j\in\mathbb{Z}\colon\beta_n\in\mathbb{R}}\beta_n^+|u_n^+ + v_n^+|^2 \geq 0.
\end{aligned}$$

Hence, we conclude by the discussion above that

$$\operatorname{Im}\langle \mathcal{M}_{\mathrm{D}}\phi,\phi\rangle \geq 0 \qquad \text{for } \phi\in H_\alpha^{-1/2}(\Gamma),$$

which means that the non-selfadjoint part $\operatorname{Im}\mathcal{M}_{\mathrm{D}}^*$ is negative semidefinite.

Theorem I.7 on range identities offers two possibilities to finish the construction of the Factorization method. Either we show that $\operatorname{Im}\mathcal{M}_{\mathrm{D}}^*$ is negative on the kernel of $\operatorname{Re}\mathcal{M}_{\mathrm{D}}^*$ or we show that $\mathcal{M}_{\mathrm{D}}^*$ is injective. The first option is difficult to show directly, but it is possible to prove that $\operatorname{Re}\mathcal{M}_{\mathrm{D}}^*$ is injective for all but a countable sequence of frequencies using holomorphic Fredholm theory. With that information, we can even apply the old version of Theorem I.7 due to Kirsch and Grinberg. On the other hand, it is much easier to show that $\mathcal{M}_{\mathrm{D}}^*$ is injective for all regular frequencies $\omega\in\mathbb{R}_{>0}\setminus\mathcal{E}_{\mathrm{d}}$. Additionally, we do not have to take further exceptional frequencies into account, which always happens when one applies holomorphic Fredholm theory. This second option is now used in the sequel. After the construction of the Factorization method in Theorem II.19 we briefly consider the first option to show how much easier the construction of the method is when one applies our modified version of the basic Theorem I.7 on range identities.

THEOREM II.19. *Let $\omega\in\mathbb{R}_{>0}\setminus\mathcal{E}_{\mathrm{d}}$ be a positive regular frequency and let $(\lambda_j,(a_n^j))_{j\in\mathbb{N}}$ be an eigensystem of $N_\sharp = |\operatorname{Re}N| + \operatorname{Im}N : \ell^2\to\ell^2$. Then a* *point $y \in \Pi$ belongs to Ω_+ if and only if the Rayleigh sequence $(g_n(y))_{n\in\mathbb{Z}}$ belongs to the range of $N_\sharp^{1/2}$, or equivalently,*

$$y \in \Omega_- \qquad \textit{if and only if} \qquad \sum_{j=1}^{\infty} \frac{|\langle (g_n(y)), (a_n^j) \rangle_{\ell^2}|^2}{\lambda_j} < \infty. \tag{II.51}$$

Proof. Theorem 3.2 in [7] shows how to consider the factorization $N = -\mathcal{L}\mathcal{M}_\mathrm{D}^*\mathcal{L}^*$ using $L^2(\Gamma)$ instead of $H_\alpha^{\pm 1/2}(\Gamma)$ as space for the middle operator. Due to Proposition II.17, $\mathcal{M}_\mathrm{i}^+$ is a coercive and selfadjoint operator between $H_\alpha^{-1/2}(\Gamma)$ and $H_\alpha^{1/2}(\Gamma)$, thus it possesses a coercive square root ι which is bounded from $H_\alpha^{-1/2}(\Gamma)$ into $L^2(\Gamma)$ and also from $L^2(\Gamma)$ into $H_\alpha^{1/2}(\Gamma)$. Moreover, ι is selfadjoint with respect the inner product of $L^2(\Gamma)$. The definition $\widetilde{\mathcal{L}} = \mathcal{L}\iota$ and $\widetilde{\mathcal{M}}_\mathrm{D} = \iota^{-1}\mathcal{M}_\mathrm{D}\iota^{-1}$ implies that the factorization

$$N = -\widetilde{\mathcal{L}}\,\widetilde{\mathcal{M}}_\mathrm{D}^*\,\widetilde{\mathcal{L}}^* \tag{II.52}$$

holds and $\widetilde{\mathcal{L}} : L^2(\Gamma) \to \ell^2$ is compact with dense range. Moreover, $\mathrm{Re}\,\widetilde{\mathcal{M}}_\mathrm{D}^* = \iota^{-1}\mathrm{Re}\,\mathcal{M}_\mathrm{D}^*\iota^{-1} : L^2(\Gamma) \to L^2(\Gamma)$ is a compact perturbation of a coercive operator and $\mathrm{Im}\,\widetilde{\mathcal{M}}_\mathrm{D}^*$ is negative semidefinite. In this situation, Theorem I.7 shows that the ranges of $N_\sharp^{1/2}$ and $\widetilde{\mathcal{L}}$ coincide, if we are able to show that $\widetilde{\mathcal{M}}_\mathrm{D}^*$ is injective. To this end, we show that $\mathcal{M}_\mathrm{D}^*$ is injective: Assume that $\mathcal{M}_\mathrm{D}^*\phi = 0$ for $\phi \in H_\alpha^{-1/2}(\Gamma)$. We observe from (II.38) that M_D^* has a non-trivial kernel, too. Since M_D^* is a Fredholm operator of index zero we conclude that also M_D has a non-trivial kernel. Then the function v defined in (II.35) solves the transmission problem (II.11) for homogeneous Dirichlet and Neumann jump data. However, our assumption that $\omega \in \mathbb{R}_{>0} \setminus \mathcal{E}_\mathrm{d}$ is a regular frequency implies that ϕ vanishes.

We established in Lemma II.11 that the range of L characterizes the interface Γ. This characterization holds also for $\widetilde{\mathcal{L}}$ from (II.52) as the ranges of $\mathcal{L}$ and L are by construction equal, see (II.37). The ranges of $\mathcal{L}$ and $\widetilde{\mathcal{L}} = \mathcal{L}\iota$ coincide, since ι is an isomorphism. Hence,

$$(g_n(y))_n \in \mathrm{Range}(\widetilde{\mathcal{L}}) \qquad \text{if and only if} \qquad y \in \Omega_-.$$

The criterion (II.51) is a consequence of Picard's range criterion. □

The last theorem strongly benefits from the generalization in Theorem I.7 on range identities compared to earlier versions of this result. To apply the ancestor of Theorem I.7, see, e.g., Theorem [49], one needs to show that the selfadjoint part $\mathrm{Re}\,\mathcal{M}_{\mathrm{D}}^*$ is injective. This can indeed be achieved via holomorphic Fredholm theory, but one has to pay the price of possibly loosing another at most countable sequence of exceptional frequencies. We briefly sketch this procedure now; our main motivation is to show the advantage of our version of Theorem I.7.

For the proof of injectivity of $\mathrm{Re}\,\mathcal{M}_{\mathrm{D}}^*$ for all but a discrete set of real frequencies we exploit the following geometric assumption:

$$\text{The interface } \Gamma \text{ is sufficiently afar from the line } \{x \in \Pi,\, x_2 = 0\}. \tag{II.53}$$

This assumption is purely technical and can always be guaranteed using a suitable coordinate transform, which is however tedious in our special context since then the incident fields in (II.32) look more complicated. The precise meaning of "sufficiently afar" will become clear in the proof of the next lemma.

LEMMA II.20. *Assume that the distance of* Γ *to the line* $\{x_2 = 0\}$ *is large enough. Then the selfadjoint part* $\mathrm{Re}\,\mathcal{M}_{\mathrm{D}}$ *is injective for* $\omega = \mathrm{i}$.

Proof. Recall that $\mathcal{M}_{\mathrm{D}} = \mathcal{M} + R$. We know from Proposition II.17 that for $\omega = \mathrm{i}$, $\mathcal{M} = \mathcal{M}_{\mathrm{i}} = \mathrm{Re}\,\mathcal{M}_{\mathrm{i}}$ is coercive on $H_\alpha^{-1/2}(\Gamma)$. The idea of the proof is now to show that the "reflection part" R is small if Γ is away from $\{x_2 = 0\}$. The operator R consists in the evaluation of trace and normal trace of single and double layer potentials, defined on the reflected interface $\Gamma' = \{x' = (x_1, x_2) : x \in \Gamma\}$, on Γ . However, for $\omega = \mathrm{i}$ these potentials and all their derivatives decay exponentially in the second variable x_2, which is easy to see from Green's function expansion (II.13). Therefore the norm of R can be made arbitrarily small by increasing the space between Γ and $\{x_2 = h\}$. If the norm of R is smaller than the coercivity constant of $\mathcal{M}$ we obtain that $\mathrm{Re}\,\mathcal{M}_{\mathrm{D}} = \mathcal{M} + \mathrm{Re}\,R$ is injective. □

Inspecting the proof of Lemma II.20, one could explicitely compute a lower bound for the minimum distance between Γ and $\{x_2 = 0\}$ in terms of the coercivity constant of $\mathcal{M}_{\mathrm{i}}$ and the exponential decay of the Green's function.

PROPOSITION II.21. *The operator* $\mathcal{M}_{\mathrm{D}}$ *depends holomorphically on the frequency* $\omega \in \mathbb{C}_+ \setminus \mathcal{E}_{\mathrm{d}}$.

Proof. Recall that $\mathcal{M}_{\mathrm{D}}$ is explicitely represented in (II.39). Using Theorem II.7 we observe that all boundary integral operators in (II.39) are holomorphic functions of the frequency. In Lemma II.9 we showed that $\Upsilon_{(\varepsilon_\pm,\pm)}$ is holomorphic in ω in $\mathbb{C}_+ \setminus \mathcal{E}_{\mathrm{d}}$. Hence follows the proposition. □

The last result together with Theorem II.3 and Lemma II.20 implies the following corollary.

COROLLARY II.22. *Under the assumptions of Lemma II.20 the selfadjoint part* $\operatorname{Re}\mathcal{M}_{\mathrm{D}}$ *is an isomorphism for all but a countable number of frequencies without finite accumulation point. The exceptional frequencies for which* $\operatorname{Re}\mathcal{M}_{\mathrm{D}}$ *fails to be injective are denoted by* $\mathcal{E}_{\mathrm{i}}$ *and we define* $\mathcal{E}_{\mathrm{t}} = \mathcal{E}_{\mathrm{d}} \cup \mathcal{E}_{\mathrm{i}}$.

By Corollary II.22, one can construct a Factorization method for the inverse periodic transmission problem in a similar way as we did it in Theorem II.19, but one can now use older results on range identities for this construction. The price one has to pay is the possible exclusion of exceptional frequencies, denoted by $\mathcal{E}_{\mathrm{i}}$ in the previous lemma, and of course a rather complicated series of arguments leading to Corollary II.22. Application of the new result on range identities in Theorem I.7 via the injectivity of the middle operator $\mathcal{M}_{\mathrm{D}}^*$ itself is much easier here.

To give a first impression of the ability of the method in reconstructing periodic interfaces, we consider a grating Γ given by the function $f(x_1) = 1.5 + 0.8\sin(x_1)$. The wavenumber k_+ above Γ equals 8.5, the wavenumber k_- is 1.5 and the quasiperiodicity α vanishes. The wavelength in the upper medium is hence 0.74. We compute the Rayleigh coefficients of the scattered fields corresponding to the incident fields from (II.32) for $j = -8, \dots, 8$ by a finite element method using two Dirichlet-to-Neumann operators on the artificial surfaces above and below the grating. Note that, due to the wavenumber 8.5, all computed modes are propagating. In other words, the reconstructions below are obtained without using evanescent modes. Using merely the propagating modes naturally deteriorates the reconstruction, see [3] for a detailed study in case of a corresponding Dirichlet problem. Nevertheless, the plots below show that one is able to get a rough estimate of the interface between the two media merely from the propagating modes. The Rayleigh coefficients which provide the data for the reconstruction are numerically computed at the top of the computational domain where $x_2 = 4$, roughly two wavelengths away from the interface. We use all 17 available Rayleigh coefficients to approximate the inner product in ℓ^2 arising in the Picard criterion and plot the reciprocal of the Picard series truncated at some truncation index. In Figure II.2 we add no artificial noise to the synthetic data and use all available terms of the Picard series in the plot. The peaks of the interface are well found by the method and in the region between $x_1 = 0$ and $x_1 = \pi$ the shape of the interface is visible in the blue part of the plot. In Figure II.3 we add one percent of artificial noise measured in the Frobenius matrix norm to the measurement data and use merely the first 10 terms of the Picard series in the plot. The first 10 terms correspond, by visual inspection, to eigenvalues which still have a reasonable precision after perturbation of the data by artificial noise. Still, the peaks of the interface are found by the method, the blue area below the maximum of the grating function is even larger than in the upper plot. However, in the area above the surface there appear periodic vertical artefacts. These artefacts become more and more pronounced if the noise level is increased.

In general, the quality of reconstructions of the Factorization method for the periodic transmission problem is not as satisfactory as the numerical examples shown in [3] for the corresponding Dirichlet problem. However, in [66] the authors report the same observation for the Linear Sampling Method applied to different inverse scattering problems involving bounded scatterers.

II-8. Remarks on the Dirichlet Scattering Problem

In [7, 3] a similar version of the Factorization method for periodic structures has been established for the inverse Dirichlet scattering problem. The authors of [7] prove a factorization of the corresponding near field operator of the form $N = -LS^*L^*$ which involves the single layer operator S on Γ, see [7, Theorem 2.2]. In the factorization of [3], S^* is replaced by the single layer operator S_{D}^* for the Dirichlet Green's function of the half space. In the two papers [7] and [3] the authors exploit positiveness of $\operatorname{Im} S^*$ and $\operatorname{Im} S_{\mathrm{D}}^*$, which does not seem to be satisfied (but semidefiniteness holds). However, this small gap in their proof can be closed in two ways: the first is to use holomorphic Fredholm theory, in the same

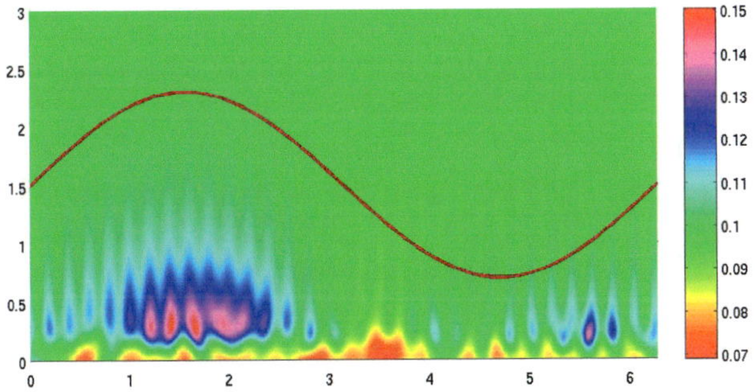

Figure II.2: Reconstruction of a 2π-periodic interface using merely propagating modes, the Picard series is truncated after 15 terms, no artificial noise.

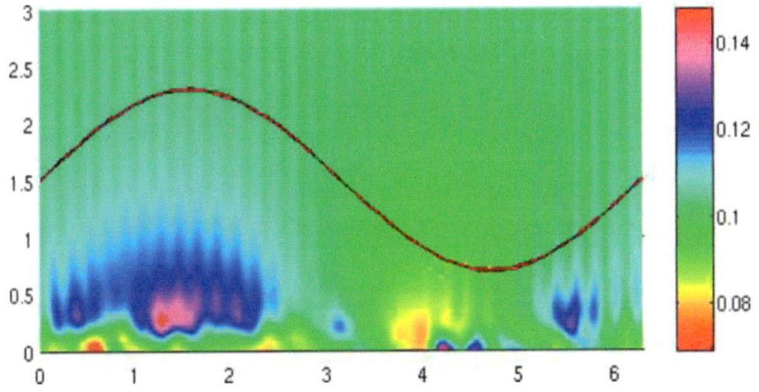

Figure II.3: Reconstruction of a 2π-periodic interface using merely propagating modes, the Picard series is truncated after 10 terms, one percent artificial noise.

fashion as above in the end of Section II-7, and to derive a result similar to Theorem II.19, valid for all but a countable set of real frequencies. We will not go into the details of that approach. The second possibility is to observe that in case that the periodic scattering surface is the graph of a function, one does not need positivity of $\operatorname{Im} S^*$ and $\operatorname{Im} S_{\mathrm{D}}^*$, since S^* and S_{D}^* are injective. Let us briefly indicate how to proceed for this approach.

For an application of Theorem I.7, we merely need to show that S and S_{D} are injective operators to establish a Factorization method for the inverse Dirichlet problem as in [7, Theorem 3.4] or [3, Theorem 2.7]. Here we note another time the advantage of Theorem I.7 compared to earlier versions of this corresponding functional analytic basis of the Factorization method. To prove injectivity of S, let us assume that there are two densities ϕ_1 and $\phi_2 \in H_\alpha^{-1/2}(\Gamma)$ such that $S\phi_1 = S\phi_2$. Then the two potentials $u_1 = \mathrm{SL}\,\phi_1$ and $u_2 = \mathrm{SL}\,\phi_2$ take the same Dirichlet boundary values on Γ. However, if Γ is the graph of a function (and sufficiently smooth), the Dirichlet scattering problem in Ω_+ together with the Rayleigh expansion condition (II.3) has a unique solution, see, e.g. [45]. The analogous scattering problem in Ω_- also has a unique solution. Therefore $u_1 = u_2$ in Π and from the jump relations for the single layer operator we obtain $\phi_1 = \phi_2$ and hence injectivity of S. The proof for S_{D} is similar, except that one has additionally to assume that $\omega^2\varepsilon$ is not a Dirichlet eigenvalue of $-\Delta$ in the domain D_-.

CHAPTER III

Detecting contamination on a rough surface

III-1. Introduction

Rough surface scattering theory in the frequency domain is a relatively new and rapidly growing field. Starting with the works of Chandler-Wilde, Ross and Zhang, see, e.g., [15, 16], on integral equation methods in two dimensions, considerable progress in understanding mathematical features arising from the unboundedness of the infinite scattering surface has been made. Inverse scattering problems for unbounded rough surfaces have on the other hand received little attention as yet. In [54], the authors investigate the detection of a contamination on a locally perturbed half-plane. In [59, 60], the authors investigate inverse scattering problems for rough surfaces using the point source method. Only recently, the Kirsch-Kress method has been investigated in the context of rough surface inverse scattering in [9]. Applications of the inverse rough surface scattering problem in non-destructive testing range from detecting impurities in industrial workpieces and inverse problems in geophysics to reconstruction problems in nanotechnology. Especially, the case of a locally perturbed periodic surface is included in our investigations. This problem occurs for instance when imperfections of a photonic crystal are searched for. Note here the link to the last chapter where we investigated the determination of a perfectly periodic photonic crystal.

In this chapter, we develop a Factorization method for the inverse scattering problem of finding a local contamination of a known infinite rough surface. Compared to Factorization methods for bounded obstacles the mathematical challenges increase, due to the different character of the radiation condition: The upward radiation condition, an analogue to the Sommerfeld radiation condition for unbounded scattering problems, yields only a semidefinite imaginary part of the middle operator of the factorization. Since we use near field measurements, the Factorization method relies on modified, namely complex conjugated, incident point sources. These sources have no physical meaning, however, it is yet an open problem to formulate a factorization of the physical measurement operator which relies on upwards propagating point sources $G(\cdot, y)$. We discuss in the second-to-last section of this chapter how to express the physically meaningfull near field operator using modified incident fields and construct a numerical algorithm for the reconstruction of the local contamination. These results extend and improve the earlier analysis for conjugate incident point sources in [48].

To give a somewhat more precise outlook on this chapter, suppose a rough surface Γ in $\mathbb{R}^2$ is locally perturbed, that is, the perturbation Γ_c differs from Γ only in a bounded set. We consider the inverse problem of reconstructing the contamination $\Gamma_c \setminus \Gamma$ using scattered waves measured on some measurement line in a finite distance above Γ. See page 105 for a rigorous mathematical statement of this inverse problem. For this problem we formulate a Factorization method which relies again on the range identity of Theorem I.7. To set the stage for the Factorization method as an inverse problems tool, we formulate the direct scattering problem using boundary integral equations on the local contamination of the rough surface. This leads in Section III-3 to a solution concept in Sobolev spaces, a necessary tool for the construction of a Factorization method. The analysis of the boundary integral operators is carried out by combination of variational methods for rough surface scattering and Green's first identity. As we have seen in Chapter III, there are quite stringent prerequisites to be satisfied for a Factorization method to be applicable. In this chapter's setting, the difficulty arising from the unboundedness of the rough surface is that the upward radiation condition makes the middle operator of the factorization non-negative, but not positive as Sommerfeld's radiation condition does. This difficulty is tackled in Section III-4 by an application of Theorem I.7, which is specifically designed for such cases.

III-2. The Unperturbed Rough Surface Scattering Problem

Time-harmonic scattering from a rough surface Γ, either of acoustic waves or of electromagnetic waves in transverse magnetic mode, compare Section I-2, can be formulated as a boundary value problem for the Helmholtz equation $\Delta u + k^2 u = 0$. The wavenumber k is in this chapter assumed to be positive and the rough surface $\Gamma = \{(x_1, f(x_1)) : x_1 \in \mathbb{R}\} \subset \mathbb{R}^2$ is the graph of a function

$$f \in C^{1,1}(\mathbb{R}) = \left\{ \phi \in C^1(\mathbb{R}) : \|\phi\|_\infty < \infty,\, \|\phi'\|_\infty < \infty, \right. \\ \left. \sup_{s \neq t \in \mathbb{R}} \frac{|\phi'(s) - \phi'(t)|}{|s - t|} < \infty \right\}.$$

For simplicity we assume that $0 < f_- < f < f_+$ and denote the domain above the surface by

$$D = \{(x_1, x_2) : x_1 \in \mathbb{R},\, x_2 > f(x_1)\} \subset \mathbb{R}^2.$$

Additionally, the line $\Gamma_h = \{(x_1, h) : x_1 \in \mathbb{R}\}$ for $h > f_+$ is often used and the strip between Γ and Γ_h is denoted by

$$D_h = \{x \in D : x_2 < h\}.$$

We also set

$$U_h = \{x \in D : x_2 > h\}$$

to be the half space above Γ_h and note that $D = D_h \cup \Gamma_h \cup U_h$. We choose the unit normal field ν on Γ to point downwards whereas on Γ_h we choose the normal ν to point upwards. Hence, ν is the exterior normal field to the domain D_h.

With the help of the fundamental solution Φ of the Helmholtz equation in two dimensions, see (I.14), we define the Dirichlet Green's function* G for the upper half space U_0,

$$G(x, y) := \Phi(x, y) - \Phi(x, y'), \qquad x \neq y,\, x_2, y_2 > 0.$$

*The quasiperiodic Green's function has been denoted by G_{D} in the last chapter, however, for simplicity we simply write G for the Dirichlet Green's function in this chapter.

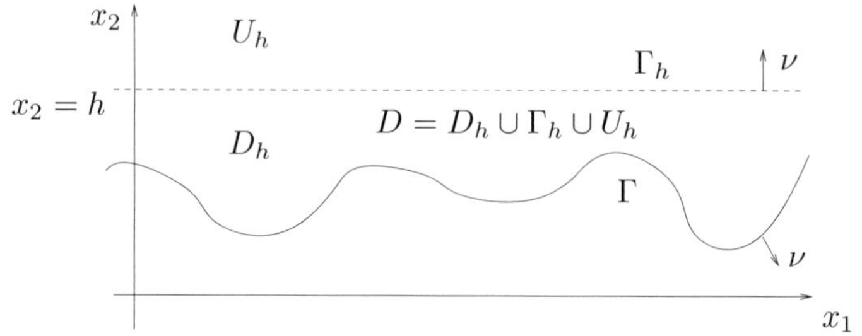

Figure III.1: The geometry of the rough surface Γ and the domain D, which is the union of the strip D_h, the line Γ_h and the half space U_h. The field ν is the exterior normal to D_h.

The point source $G(\cdot, y)$, $y \in D$, satisfies the Helmholtz equation in $D \setminus \{y\}$. We use such functions as incident fields, $u^i(\cdot, y) := G(\cdot, y)$. The corresponding total field u needs to satisfy the sound-soft boundary condition $u = 0$ on Γ and, as usual in time harmonic scattering, the scattered field $u^s = u - u^i$ needs to satisfy a radiation condition, which is for our geometry the upward radiation condition [15]. In the following, we formulate the scattering problem for the scattered field u^s with general Dirichlet boundary values in the space $BC(\Gamma)$ of bounded and continuous functions on Γ: Given $g \in BC(\Gamma)$, find $u^s \in C^2(D) \cap C(\overline{D})$ such that

$$\Delta u^s + k^2 u^s = 0 \text{ in } D, \qquad u^s = g \text{ on } \Gamma, \tag{III.1}$$

and u^s satisfies the following upward radiation condition: For some $h > f_+$ and $\phi \in L^\infty(\Gamma_h)$,

$$u^s(x) = \int_{\Gamma_h} \frac{\partial \Phi}{\partial y_2}(x, y)\phi(y)\,\mathrm{d}s(y), \qquad x \in U_h. \tag{III.2}$$

A solution to this problem is found in form of a Brakhage-Werner-type ansatz,

$$u^s(x) = \int_\Gamma \left(\frac{\partial G}{\partial \nu(y)}(x, y) - i\eta G(x, y) \right) \psi(y)\,\mathrm{d}s(y), \tag{III.3}$$

with a density $\psi \in BC(\Gamma)$, $\eta > 0$ fixed and $x \in D$. This problem is well known to be equivalent to the boundary integral equation

$$\psi + 2\int_\Gamma \frac{\partial G}{\partial \nu(y)}(\cdot, y)\psi(y)\,\mathrm{d}s(y) - 2i\eta \int_\Gamma G(\cdot, y)\psi(y)\,\mathrm{d}s(y) = -2g \quad \text{(III.4)}$$

which has been proved to be uniquely solvable [5, Corollary 4.5]. From now on we restrict ourselves to the case $g = G(\cdot, y)$, $y \in D$, denote by $u^s(\cdot, y)$ the corresponding solution of (III.1) and set $u(\cdot, y) = G(\cdot, y) + u^s(\cdot, y)$.

Let us recall several regularity and decay results for u^s and u. In [16, Theorem 3.1], see also [5, Section 4], it is shown that for $h > f_+$ and $\alpha \in (0, 1)$ there exists $C > 0$ such that

$$|\nabla u(x)| \leq C(x_2 - f(x_1))^{\alpha-1}, \qquad x \in D_h. \quad \text{(III.5)}$$

The bound (III.5) can be used to prove a Green's representation theorem similar as in [16, Theorem 3.3] or [85, Theorem 4.1],

$$u^s(x, y) = \int_\Gamma \left(G(x, z)\frac{\partial u^s}{\partial \nu(z)}(z, y) - u^s(z, y)\frac{\partial G}{\partial \nu(z)}(x, z) \right) \mathrm{d}s(z), \quad \text{(III.6)}$$

for $x, y \in D$. Here, ν denotes the exterior unit normal field to D. Note that a similar representation where the kernel G is replaced by the free space Green's function Φ does not hold. Results from [5, 4] can be used to obtain decay rates for the scattered field u^s as follows. From [15] we know that

$$|G(x, y)|, |\nabla_y G(x, y)| \leq C\frac{(1 + x_2)(1 + y_2)}{1 + |x - y|^{3/2}}, \qquad x, y \in D, |x - y| \geq 1, \quad \text{(III.7)}$$

and therefore we obtain from Corollary 4.5 in [5] that the density ψ from (III.4) decays as $|x_1|^{-3/2}$. The same arguments as in the proof of [4, Theorem 4.6, Corollary 4.7] show that u^s, defined in (III.3), has the same decay rate in x_1 in the strip D_h,

$$|u^s(x, y)| \leq C|x_1 - y_1|^{-3/2}, \qquad x \in D_h, |x_1 - y_1| > 1. \quad \text{(III.8)}$$

The bound (III.5) implies that ∇u is locally square integrable in $\overline{D}$. Thus, for any ball $B(x, r)$ of radius $r > 0$ around $x \in \Gamma$, $\|u\|_{H^1(D\cap B(x,r))}$ is finite, at least if the singularity of the incident field $G(\cdot, y)$ at y is far enough away from x. Elliptic regularity results [28, Corollary 8.36] state that there is a constant $C(x)$ such that for any ball $B(x,r')$ of radius $r' < r$ around x

$$\|u(\cdot,y)\|_{C^{1,\alpha}(\overline{D\cap B(x,r')})} \le C(x)\|u(\cdot,y)\|_{C^0(\overline{D\cup B(x,r)})}.$$

The constant $C(x)$ depends on the local Hölder norm of the function f which defines Γ. As $\|f\|_{C^{1,\alpha}(\mathbb{R})}$ is bounded (we assumed $f \in C^{1,1}(\mathbb{R})$), the set $\{C(x) : x \in \Gamma\}$ is uniformly bounded. Then the decay of $u(\cdot,y) = G(\cdot,y) + u^s(\cdot,y)$ implies

$$\|u(\cdot,y)\|_{C^{1,\alpha}(D\cap B(x,r))} \le C|x_1 - y_1|^{-3/2}, \quad x \in D_h,\ |x_1 - y_1| > 1$$

that is, the decay of the total field u carries over to ∇u. From the rapid decay of the Green's function in (III.7) we conclude that u^s belongs to $H^1(D_h)$ and $u(\cdot,y)$ belongs to $H^1(D_h \setminus B(y,1))$ for $h > f_+$.

The smoothness assumptions on Γ are not the weakest possible. Using results and techniques from [16, Theorem 3.1] one could also deal with piecewise Lyapunov surfaces, however, we restrict ourselves to the class $C^{1,1}$ to avoid technical complications.

In the analysis of the Factorization method for the inverse scattering problem we make use several times of the following reciprocity lemma, which shows that the Green's function for D is symmetric. See also [59, 60] for a proof.

LEMMA III.1. *Denote by $u^s(\cdot,y)$ the scattered field corresponding to the incident field $G(\cdot,y)$. Then the total field $u = u^s(\cdot,y) + G(\cdot,y)$ satisfies the reciprocity relation $u(x,y) = u(y,x)$ for $x \neq y \in D$.*

Proof. By the symmetry of the Green's function G,

$$u^s(x,y) = \int_\Gamma \left(G(z,x)\frac{\partial u^s}{\partial \nu(z)}(z,y) - u^s(z,y)\frac{\partial G}{\partial \nu(z)}(z,x) \right) \mathrm{d}s(z). \quad \text{(III.9)}$$

Interchanging the roles of x and y yields, for $x, y \in D$,

$$u^s(y,x) = \int_\Gamma \left(G(z,y)\frac{\partial u^s}{\partial \nu(z)}(z,x) - u^s(z,x)\frac{\partial G}{\partial \nu(z)}(z,y) \right) \mathrm{d}s(z). \quad \text{(III.10)}$$

Let $h > f_+$ be such that u^s satisfies the upward radiation condion in (III.2). By Green's second identity applied in the interior of the contour

$\partial D_{a,b}$ from Figure III.2 for $b > 0$ large enough and $a > h$,

$$0 = \int_{\partial D_{a,b}} \left(u^s(z,x) \frac{\partial u^s}{\partial \nu(z)}(z,y) - u^s(z,y) \frac{\partial u^s}{\partial \nu(z)}(z,x) \right) \mathrm{d}s(z). \quad \text{(III.11)}$$

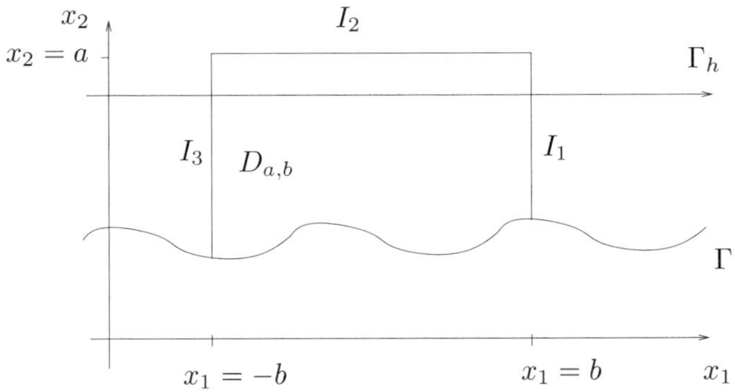

Figure III.2: The contour $D_{a,b}$.

Let us first show that the part of the integral on the vertical line $I_1 := \{x \in D : x_1 = b, x_2 < a\}$ tends to zero as $b \to \infty$. It is sufficient to consider

$$\int_{I_1} u^s(z,x) \frac{\partial u^s}{\partial \nu(z)}(z,y) \, \mathrm{d}s(z) \qquad \text{as } b \to \infty.$$

From the boundedness condition of the scattering problem (III.1) and estimate (III.5) we observe that $\int_{I_1} |\partial u^s / \partial \nu| \, \mathrm{d}s(z)$ is well defined and uniformly bounded in $b > 0$. By the decay rate (III.8) it holds

$$\left| \int_{I_1} u^s(z,x) \frac{\partial u^s}{\partial \nu(z)}(z,y) \, \mathrm{d}s(z) \right| \leq \sup_{z \in I_1} |u^s(z,x)| \int_{I_1} \left| \frac{\partial u^s}{\partial \nu(z)}(z,y) \right| \mathrm{d}s(z) \to 0$$

as $b \to \infty$ and the claim follows. Hence, taking the limit in (III.11) as $b \to \infty$ we know that

$$0 = \int_{\Gamma \cup \Gamma_a} \left(u^s(z,x) \frac{\partial u^s}{\partial \nu(z)}(z,y) - u^s(z,y) \frac{\partial u^s}{\partial \nu(z)}(z,x) \right) \mathrm{d}s(z),$$

where the integral exists due to the decay of u^s in (III.8). Additionally, the part of the integral in the latter equation on the upper line Γ_a vanishes by Theorem 2.9 from [16]. Hence we arrive at

$$0 = \int_\Gamma \left(u^s(z,x) \frac{\partial u^s}{\partial \nu(z)}(z,y) - u^s(z,y) \frac{\partial u^s}{\partial \nu(z)}(z,x) \right) \, \mathrm{d}s(z). \qquad \text{(III.12)}$$

A similar formula for the incident fields fails due to the singularity of the field: we can only use Green's formula on $D_{a,b} \setminus (B(x,\varepsilon) \cup B(y,\varepsilon))$ for $\varepsilon > 0$ small enough, which yields by similar arguments as above

$$\begin{aligned} 0 = &\int_\Gamma \left(G(z,x) \frac{\partial G}{\partial \nu(z)}(z,y) - G(z,y) \frac{\partial G(z,x)}{\partial \nu(z)} \right) \, \mathrm{d}s(z) \\ &+ \int_{\partial B(x,\varepsilon) \cup \partial B(y,\varepsilon)} \left(G(z,x) \frac{\partial G}{\partial \nu(z)}(z,y) - G(z,y) \frac{\partial G}{\partial \nu(z)}(z,x) \right) \, \mathrm{d}s(z). \end{aligned}$$

As $\varepsilon \to 0$ we obtain as in the proof of [22, Theorem 2.1]

$$G(x,y) - G(y,x) = \int_\Gamma \left(G(z,x) \frac{\partial G}{\partial \nu(z)}(z,y) - G(z,y) \frac{\partial G}{\partial \nu(z)}(z,x) \right) \, \mathrm{d}s(z).$$

We add the last equation to (III.9), (III.10), (III.12) and obtain

$$\begin{aligned} u^s(x,y) - u^s(y,x) = &\int_\Gamma \left(u(z,x) \frac{\partial u}{\partial \nu(z)}(z,y) - u(z,y) \frac{\partial u}{\partial \nu(z)}(z,x) \right) \, \mathrm{d}s(z) \\ &+ G(y,x) - G(x,y), \end{aligned}$$

and the boundary condition $u = 0$ on Γ yields $u^s(x,y) - u^s(y,x) = G(y,x) - G(x,y)$ and $u(x,y) = u(y,x)$. □

III-3. Local Contamination and Boundary Integral Operators

We consider in this section the direct scattering problem when the rough surface Γ is locally perturbed. The aim of this section is to prepare the tools for the treatment of the inverse problem later on, which consists in the reconstruction of the perturbation when the rough surface is known. Let us denote the perturbed surface by Γ_c and the perturbation itself by

$\mathcal{C} := \Gamma_c \setminus \Gamma$. We require that the domain between $\mathcal{C}$ and Γ, called D_-, is a Lipschitz domain contained in $D = \{(x_1, x_2) : x_1 \in \mathbb{R}, x_2 > f(x_1)\}$, where the function f defines the (unperturbed) rough surface. This assumption forbids for instance cusps at the boundary points of $\mathcal{C}$. Considering only local perturbations means that we assume that D_- is bounded. The complement of $\overline{D_-}$ in D is called D_+ and we set

$$D_h^+ = D_h \cap D_+ = \{(x_1, x_2) : x_1 \in \mathbb{R}, f(x_1) < x_2 < h\} \cap D_+,$$

where h is now and in the following assumed to be so large that Γ_h and $\mathcal{C}$ do not intersect. Moreover, we suppose that D_+ satisfies the *secondary geometric assumption*, compare [14]:

$$\text{for } x = (x_1, x_2) \in D_+ \text{ and } s > 0 \text{ it holds } (x_1, x_2 + s) \in D_+. \qquad \text{(III.13)}$$

The unit normal ν on the surface Γ_c points again downwards, as the normal on Γ points downwards. Note that ν is hence the interior normal to D_- on the arc $\mathcal{C}$, but the exterior normal on $\partial D_- \cap \Gamma$.

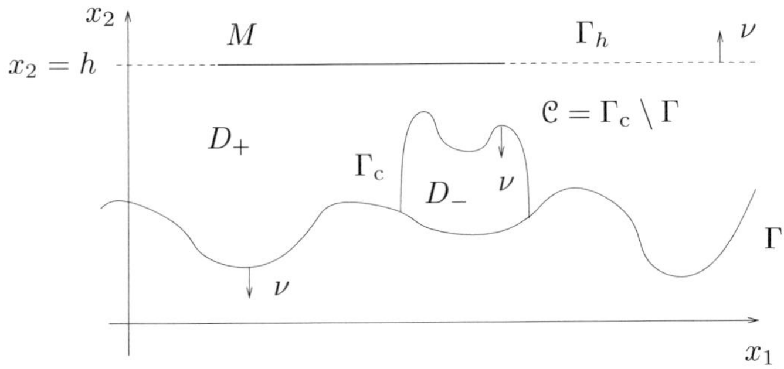

Figure III.3: The unperturbed surface Γ and the perturbed surface Γ_c. The unit normal ν to Γ and Γ_c points downwards. On the measurement line M, to be introduced in Section III-4, data for the inverse problem is collected.

We consider again incident point sources $G(\cdot, y)$ for $y \in D_+$ and represent the total field u by

$$u(\cdot, y) = G(\cdot, y) + u^s(\cdot, y) + u^c(\cdot, y), \qquad x,\, y \in D_+, \qquad \text{(III.14)}$$

where $u^s(\cdot, y)$ is the scattered field corresponding to the incident field $G(\cdot, y)$ scattered at the surface Γ. Moreover, $u^c(\cdot, y)$ is the part of the total field which corresponds to scattering by the contamination $\mathcal{C}$: u^c is an upwards radiating solution of the Helmholtz equation in D_+ that satisfies the boundary condition

$$u^c(\cdot, y) = -\big(G(\cdot, y) + u^s(\cdot, y)\big) \qquad \text{on } \Gamma_c.$$

We note that the boundary values $-\ (G(\cdot, y) + u^s(\cdot, y))|_{\Gamma_c}$ have compact support. Therefore we find u^c by an integral equation approach on the contamination $\mathcal{C}$, leading to a solution concept in Sobolev spaces. The Sobolev space setting is important for the application of the Factorization method later on.

The boundary ∂D_- is the extension of the curve $\mathcal{C}$ by a part of Γ to a closed Lipschitz curve. Following [80], see also [50, 53, 63], we set

$$\widetilde{H}^s(\mathcal{C}) := \left\{ \phi|_{\mathcal{C}} : \phi \in H^s(\partial D_-),\ \operatorname{supp}(\phi) \subset \overline{\mathcal{C}} \right\}, \qquad |s| \le 1,$$

and

$$H^s(\mathcal{C}) := \{ \phi|_{\mathcal{C}} : \phi \in H^s(\partial D_-)\}, \qquad |s| \le 1.$$

We recall that $\widetilde{H}^s(\mathcal{C})$ is also the completion of $C_c^\infty(\mathcal{C})$ in the norm of $H^s(\mathcal{C})$. The couples $\langle \widetilde{H}^{-s}(\mathcal{C}), H^s(\mathcal{C})\rangle$ and $\langle H^{-s}(\mathcal{C}), \widetilde{H}^s(\mathcal{C})\rangle$ are dual pairings for the inner product of $L^2(\mathcal{C})$ and form a Gelfand triple [63]. In contrast to [50, 80, 53], we are going to work with the pair $\langle H^{-s}(\mathcal{C}), \widetilde{H}^s(\mathcal{C})\rangle$, since in our case the Dirichlet values of $u^c = -(G(\cdot, y) + u^s(\cdot, y))$ on $\mathcal{C}$ vanish in the endpoints of $\mathcal{C}$. Obviously, the formulation of the scattering problem gets easier if the local perturbation does not involve the boundary Γ of the domain, that is, when $\overline{D_-}$ and Γ do not intersect. Then $\widetilde{H}^s(\mathcal{C}) = H^s(\mathcal{C})$.

In the Sobolev space setting, the boundary value problem for u^c is formulated as follows: Given $g \in \widetilde{H}^{1/2}(\mathcal{C})$, find $u^c \in H^1_{\text{loc}}(D_+)$ such that

$$\Delta u^c + k^2 u^c = 0 \text{ in } D_+, \qquad u^c = g \text{ on } \mathcal{C}, \qquad u^c = 0 \text{ on } \Gamma_c \setminus \mathcal{C}, \quad \text{(III.15)}$$

the norm $\|u^c\|_{H^1(D_h^+)}$ is bounded and u^c satisfies the upward radiation condition (III.2).

Problem (III.15) is of course a special case of a rough surface scattering problem in the domain D_+. In the following, we collect results from [14] on variational methods for such problems. These results are subsequently applied to show boundedness of the single layer operator on $\mathcal{C}$ which we will introduce to deal with the direct scattering problem (III.15). The idea of proving properties of boundary integral operators via properties of corresponding variational problems is well known, and has been also applied for instance in the monograph [63], but also in [80]. In the latter reference, the authors treat scattering problems for screens and their results are in some sense analogous to ours.

Scattering problems for the Helmholtz equation in D_h^+ can be cast into variational formulations using Green's first identity. In case that $u \in H^1(D_h^+)$ solves $\Delta u + k^2 u = f$ with $f \in L^2(D_h^+)$ and satisfies as well the boundary condition $u|_{\Gamma_c} = 0$, Green's first identity shows that

$$\int_{D_h^+} \left(\nabla u \nabla \overline{v} - k^2 u \overline{v}\right) \mathrm{d}x + \int_{D_h^+} f \overline{v} \,\mathrm{d}x = \int_{\Gamma_h} \frac{\partial u}{\partial \nu} \overline{v} \,\mathrm{d}s, \qquad \text{(III.16)}$$

first for all $v \in C_c^\infty(D_h^+)$, and hence by density for all $v \in H_0^1(D_h^+) := \{v \in H^1(D_h^+) : v|_{\Gamma_c} = 0\}$. Now we replace the normal derivative of u on Γ_h by the Dirichlet-to-Neumann operator T applied to $u|_{\Gamma_h}$, compare [14]. The Fourier transform [63]

$$\mathcal{F}\phi(\xi) = \frac{1}{\sqrt{2\pi}} \int_{-\infty}^{\infty} \phi(x) \exp(-\mathrm{i}x\xi) \,\mathrm{d}x, \qquad \xi \in \mathbb{R},$$

and Fourier's inversion formula yield

$$u(x_1, h) = \frac{1}{\sqrt{2\pi}} \int_{\mathbb{R}} \mathcal{F}u(\cdot, h)(\xi) \exp(\mathrm{i}x_1\xi) \,\mathrm{d}\xi, \qquad x_1 \in \mathbb{R}.$$

As u solves the Helmholtz equation in U_h, we formally obtain

$$u(x_1, x_2) = \frac{1}{\sqrt{2\pi}} \int_{\mathbb{R}} \mathcal{F}u(\cdot, h)(\xi) \exp\left(\mathrm{i}x_1\xi + \mathrm{i}\sqrt{k^2 - \xi^2}(x_2 - h)\right) \mathrm{d}\xi, \quad x_1 \in \mathbb{R}, \qquad \text{(III.17)}$$

and by our choice of the plus sign in front of the square one can show [14, 6] that such u satisfies the upward radiation condition (III.2). Formally taking the derivative with respect to x_2 of the latter expression yields

$$\frac{\partial u}{\partial \nu}(x_1, h) = \frac{\mathrm{i}}{\sqrt{2\pi}} \int_{\mathbb{R}} \sqrt{k^2 - \xi^2}\, \mathcal{F}u(\cdot, h)(\xi) \exp(\mathrm{i}x_1\xi) \,\mathrm{d}\xi, \quad x_1 \in \mathbb{R}, \; x_2 > h.$$

All these formal steps are valid if the involved functions are smooth enough and sufficiently decaying [14, Lemma 2.2]. Moreover, the operator T, defined by

$$(T\phi)(x) = \frac{\mathrm{i}}{\sqrt{2\pi}} \int_{\mathbb{R}} \sqrt{k^2 - \xi^2}\, \mathcal{F}\phi(\cdot, h)(\xi) \exp(\mathrm{i}x_1\xi)\, \mathrm{d}\xi, \qquad x \in \Gamma_h, \tag{III.18}$$

is bounded from $H^{1/2}(\Gamma_h)$ into $H^{-1/2}(\Gamma_h)$ [14, Lemma 2.4]. By a density argument, see [14, Lemma 2.2] for details, T is hence the Dirichlet-to-Neumann operator on Γ_h and maps a function in $\phi \in H^{1/2}(\Gamma_h)$ to the Neumann boundary values of the upwards radiating function which takes Dirichlet boundary values ϕ. Consequently, we can replace $\partial u/\partial \nu$ in (III.16) by $T(u|_{\Gamma_h})$ and obtain the following variational formulation for the scattering problem: Find $u \in H^1_0(D_h^+)$ such that

$$\mathcal{B}(u, v) := \int_{D_h^+} (\nabla u \nabla \overline{v} - k u \overline{v})\, \mathrm{d}x - \int_{\Gamma_h} \overline{v} T(u)\, \mathrm{d}s = -\int_{D_h^+} f \overline{v}\, \mathrm{d}x \tag{III.19}$$

for all $v \in H^1_0(D_h^+)$.

THEOREM III.2 (Theorem 4.1 in [14]). *The variational problem to find* $u \in H^1_0(D_h^+)$ *such that*

$$\mathcal{B}(u, v) = f(v) \qquad \textit{for all } v \in H^1_0(D_h^+)$$

has a unique solution for all continuous antilinear forms f in the dual space $H^1_0(D_h^+)^*$ of $H^1_0(D_h^+)$. This solution depends continuously on f,

$$\|u\|_{H^1(D_h^+)} \leq C \|f\|_{H^1_0(D_h^+)^*}.$$

for all $u \in H^1_0(D_h^+)$.

A well known technique allows to solve inhomogeneous rough surface Dirichlet problems.

COROLLARY III.3. *There is a constant $C > 0$ such that for each $\psi \in H^{1/2}(\Gamma_\mathrm{c})$ there is a unique solution $u \in H^1(D_h^+)$ of the Dirichlet problem*

$$\Delta u + k^2 u = 0 \quad \textit{in } D_h^+, \qquad u|_{\Gamma_\mathrm{c}} = \psi, \qquad \left.\frac{\partial u}{\partial \nu}\right|_{\Gamma_h} = T(u),$$

which satisfies $\|u\|_{H^1(D_h^+)} \leq C \|\psi\|_{H^{1/2}(\Gamma_\mathrm{c})}$. Especially, for every $g \in \widetilde{H}^{1/2}(\mathcal{C})$ there is a unique solution of problem (III.15) *in $H^1(D_h^+)$ which depends continuously on g.*

Proof. The proof follows the idea of the proof of Theorem II.5. Theorem 3.37 in [63] states that the trace operator has a continuous right inverse and hence there is $\widetilde{g} \in H^1(D_h^+)$ such that $\widetilde{g}|_{\Gamma_c} = g$ and $\|\widetilde{g}\|_{H^1(D_h^+)} \leq C\|g\|_{H^{1/2}(\Gamma_c)}$ with C independent of g. The variational problem to find $w \in H_0^1(D_h^+)$ such that

$$\mathcal{B}(w, v) = -\int_{D_h^+} (\nabla\widetilde{g}\nabla\overline{v} - k^2\widetilde{g}\overline{v})\,\mathrm{d}x \qquad \text{for all } v \in H_0^1(D_h^+)$$

is uniquely solvable according to the previous theorem. Then the sum $w + \widetilde{g}|_{D_h^+}$ is the solution we are looking for and the estimate $\|w+\widetilde{g}\|_{H^1(D_h^+)} \leq \|w\|_{H^1(D_h^+)} + \|\widetilde{g}\|_{H^1(D_h^+)} \leq C\|\widetilde{g}\|_{H^1(D_h^+)} + C\|g\|_{H^{1/2}(\Gamma_c)} \leq C\|g\|_{H^{1/2}(\Gamma_c)}$ holds. □

We also need the following corollary on transmission problems posed on the arc $\mathcal{C}$. Therefore we introduce the jump of a function u across $\mathcal{C}$ as the difference of the trace $u|_{\mathcal{C}}^+$ from D_+ and the trace $u|_{\mathcal{C}}^-$ from D_-,

$$[u]_{\mathcal{C}} = u|_{\mathcal{C}}^+ - u|_{\mathcal{C}}^- \,.$$

Corollary III.4. *For each $\phi \in H^{-1/2}(\mathcal{C})$ there is a unique solution $u \in H_0^1(D_h^+ \cup D_-) := \{u \in H^1(D_h^+ \cup D_-) : u|_\Gamma = 0\}$ of the transmission problem*

$$\Delta u + k^2 u = 0 \quad \text{in } D_h^+ \cup D_-, \quad [u]_{\mathcal{C}} = 0, \quad \left[\frac{\partial u}{\partial \nu}\right]_{\mathcal{C}} = \phi, \quad \left.\frac{\partial u}{\partial \nu}\right|_{\Gamma_h} = T(u),$$

which satisfies $\|u\|_{H^1(D_h^+ \cup D_-)} \leq C\|\phi\|_{H^{-1/2}(\mathcal{C})}$ for some constant $C > 0$.

Proof. First, we note that the domain D is a domain with Lipschitz boundary which satisfies by assumption the secondary geometric assumption (III.13). Therefore the claim of Theorem III.2 holds also for the domain D_h instead of D_h^+: The variational formulation to find u in $H_0^1(D_h) := \{u \in H^1(D_h) : u|_\Gamma = 0\}$ such that

$$\int_{D_h} (\nabla u \nabla\overline{v} - ku\overline{v})\,\mathrm{d}x - \int_{\Gamma_h} \overline{v}T(u)\,\mathrm{d}s = f(v) \tag{III.20}$$

for all $v \in H^1_0(D_h)$ has a unique solution for any continuous antilinear form f on $H^1_0(D_h)$. We will now use the form

$$f(v) = \int_{\mathcal{C}} \phi \overline{v} \, \mathrm{d}s,$$

which is bounded since $\phi \in H^{-1/2}(\mathcal{C})$ and $v|_{\mathcal{C}} \in \widetilde{H}^{1/2}(\mathcal{C})$. Our claim is that the solution for this right hand side solves the transmission problem stated in the corollary and vice versa. Consider any solution u of the transmission problem stated in the corollary. Continuity of u across the interface $\mathcal{C}$ implies that $u \in H^1_0(D_h)$, see, e.g., [63, Excercise 4.5]. Further, due to Green's first identity in the domain D_h^+ applied to $\Delta u + k^2 u$ and $\overline{v} \in H^1_0(D_h)$, the restriction $u|_{D_h^+}$ solves

$$\int_{D_h^+} (\nabla u \nabla \overline{v} - k u \overline{v}) \, \mathrm{d}x - \int_{\Gamma_h} \overline{v} T(u) \, \mathrm{d}s = \int_{\mathcal{C}} \left. \frac{\partial u}{\partial \nu} \right|_{\mathcal{C}}^{+} \overline{v} \, \mathrm{d}s$$

for all $v \in H^1_0(D_h)$, whereas $u|_{D_-}$ solves

$$\int_{D_-} (\nabla u \nabla \overline{v} - k u \overline{v}) \, \mathrm{d}x = - \int_{\mathcal{C}} \left. \frac{\partial u}{\partial \nu} \right|_{\mathcal{C}}^{-} \overline{v} \, \mathrm{d}s$$

for all $v \in H^1_0(D_h)$. Adding the latter two variational equations shows that $u \in H^1_0(D_h)$ indeed solves the variational problem (III.20). The proof that any solution to the variational problem gives rise to a solution of the transmission problem stated in the corollary uses analogous arguments. □

With our knowledge on solvability of Dirichlet and transmission problems for rough surfaces, we can now introduce the single layer potential and the single layer operator on the contamination $\mathcal{C}$. For $\phi \in H^{-1/2}(\mathcal{C})$ the single layer potential is defined as

$$(\mathrm{SL}\, \phi)(x) = \int_{\mathcal{C}} w(x, y) \phi(y) \, \mathrm{d}s(y), \qquad x \in D_+ \cup D_-, \tag{III.21}$$

with kernel

$$w(x, y) = G(x, y) + u^s(x, y), \qquad x, y \in D, \, x \neq y.$$

The kernel w is the Dirichlet Green's function for D and it is symmetric by Lemma III.1. Since $w(x,\cdot) \in \widetilde{H}^{1/2}(\mathcal{C})$, the expression in (III.21) is well defined for $x \in D_+ \cup D_-$ and it solves the Helmholtz equation in this domain. Additionally, the trace of $\mathrm{SL}\,\phi$ on Γ vanishes, as w vanishes on Γ, and $\mathrm{SL}\,\phi$ is by construction an upwards radiating function. For $x \in \mathcal{C}$ we formally define the single layer operator by restriction to $\mathcal{C}$,

$$(S\phi)(x) = \int_{\mathcal{C}} w(x,y)\phi(y)\,\mathrm{d}s(y), \qquad x \in \mathcal{C}.$$

In our next theorem, we show that this single layer operator is bounded between suitably chosen Sobolev spaces. Second, in order that the single layer potential $\mathrm{SL}\,\phi$ takes prescribed boundary values $g \in \widetilde{H}^{1/2}(\mathcal{C})$, the density $\phi \in H^{-1/2}(\Gamma)$ needs to satisfy the following boundary integral equation in $\widetilde{H}^{1/2}(\mathcal{C})$,

$$S\phi = g, \quad \text{that is,} \quad \int_{\mathcal{C}} w(\cdot,y)\phi(y)\,\mathrm{d}s(y) = g. \tag{III.22}$$

THEOREM III.5. *Assume that k^2 is not a Dirichlet eigenvalue of $-\Delta$ in the domain D_-.*

(a) The single layer operator S is bounded from $H^{-1/2}(\mathcal{C})$ into $\widetilde{H}^{1/2}(\mathcal{C})$.

(b) A function $u^c \in H^1_{\mathrm{loc}}(D_h^+)$ is a solution of (III.15) *for $g \in \widetilde{H}^{1/2}(\mathcal{C})$ if and only if*

$$u^c(x) = \int_{\mathcal{C}} w(x,y)\phi(y)\,\mathrm{d}s(y), \qquad x \in D_h^+, \tag{III.23}$$

and $\phi \in H^{-1/2}(\mathcal{C})$ satisfies the boundary integral equation (III.22).

Proof. Assume u^c is the solution of (III.15), known to exist by Corollary III.3, for some $g \in \widetilde{H}^{1/2}(\mathcal{C})$. We show that it has the form of (III.23) and obtain boundedness of the single layer operator as a side product. Recalling the definition of $D_{a,b}$ (see Figure III.2) we set $D_{a,b}^+ := D_{a,b} \cap D_+$. By Green's formula in $D_{a,b}^+$,

$$u^c(x) = \int_{\partial D_{a,b}^+} \left(G(x,y)\frac{\partial u^c}{\partial \nu}(y) - u^c(y)\frac{\partial G}{\partial \nu(y)}(x,y) \right) \mathrm{d}s(y), \quad x \in D_{a,b}^+. \tag{III.24}$$

As in the proof of Lemma III.1, this integral is split up in four parts on the lines I_1, I_2, I_3 and a part on Γ_c. The integral $\int_{I_1} |\partial u^c/\partial\nu| \, \mathrm{d}s(y)$ is uniformly bounded for all sufficiently large b since $u^c \in H^1(D_a^+)$ and therefore the part of the integral in (III.24) on I_1 tends to zero as $b \to \infty$, as well as the part on I_3. Then the part on the entire horizontal line Γ_a vanishes, again by [16, Theorem 2.9]. Hence,

$$u^c(x) = \int_{\Gamma_c} \left(G(x,y) \frac{\partial u^c}{\partial \nu}(y) - u^c(y) \frac{\partial G}{\partial \nu(y)}(x,y) \right) \mathrm{d}s(y), \quad x \in D_+. \tag{III.25}$$

Note that in this formula, ν denotes the exterior normal on D_+ pointing downwards. Let us now extend u^c to all of D by the solution of the interior Dirichlet problem

$$\Delta u^c + k^2 u^c = 0 \text{ in } D_-, \qquad u^c|_{\mathcal{C}} = g, \qquad u^c|_{\partial D_- \setminus \mathcal{C}} = 0.$$

Our assumption that k^2 is no eigenvalue of the Dirichlet problem in D_- implies that this problem has a unique solution in $H^1(D_-)$. Moreover, since $[u^c]_{\mathcal{C}} = 0$ the extension of u^c belongs to $H^1_{\mathrm{loc}}(D)$. Another application of Green's formula yields

$$0 = \int_{\partial D_-} \left(G(x,y) \frac{\partial u^c}{\partial \nu_-}(y) - u^c(y) \frac{\partial G(x,y)}{\partial \nu_-(y)} \right) \mathrm{d}s(y), \quad x \in D_+. \tag{III.26}$$

Here, ν_- denotes the exterior normal to D_-, thus $\nu_- = -\nu$ on $\mathcal{C}$. Adding equation (III.25) and (III.26) yields

$$\begin{aligned} u^c(x) &= \int_\Gamma G(x,y) \frac{\partial u^c}{\partial \nu}(y) \, \mathrm{d}s(y) + \int_{\mathcal{C}} G(x,y) \left[\frac{\partial u^c}{\partial \nu}(y) \right]_{\mathcal{C}} \mathrm{d}s(y) \\ &= - \int_\Gamma u^s(x,y) \frac{\partial u^c}{\partial \nu}(y) \, \mathrm{d}s(y) - \int_{\mathcal{C}} u^s(x,y) \left[\frac{\partial u^c}{\partial \nu}(y) \right]_{\mathcal{C}} \mathrm{d}s(y) \\ &\qquad + \int_{\mathcal{C}} w(x,y) \left[\frac{\partial u^c}{\partial \nu}(y) \right]_{\mathcal{C}} \mathrm{d}s(y), \qquad x \in D_+. \end{aligned}$$

It remains to prove that the term

$$(*) := \int_\Gamma u^s(x,y) \frac{\partial u^c}{\partial \nu}(y) \, \mathrm{d}s(y) + \int_{\mathcal{C}} u^s(x,y) \left[\frac{\partial u^c}{\partial \nu}(y) \right]_{\mathcal{C}} \mathrm{d}s(y)$$

vanishes. First we integrate by parts to obtain

$$\int_\Gamma u^s(x,y)\frac{\partial u^c}{\partial\nu}(y)\,\mathrm{d}s(y) + \int_{\mathcal{C}} u^s(x,y)\,\frac{\partial u^c}{\partial\nu}(y)\bigg|^+ \mathrm{d}s(y)$$

$$= \int_{\Gamma_\mathrm{c}} u^s(x,y)\,\frac{\partial u^c}{\partial\nu}(y)\bigg|^+ \mathrm{d}s(y) + \int_{\partial D_-\setminus\mathcal{C}} u^s(x,y)\,\frac{\partial u^c}{\partial\nu}(y)\bigg|^- \mathrm{d}s(y)$$

$$= \int_{\mathcal{C}} \frac{\partial u^s(x,y)}{\partial\nu(y)}u^c(y)\,\mathrm{d}s(y) + \int_{\partial D_-\setminus\mathcal{C}} u^s(x,y)\frac{\partial u^c}{\partial\nu}(y)\bigg|^- \mathrm{d}s(y)$$

since u^c vanishes on $\Gamma_\mathrm{c}\setminus\mathcal{C}$. Now we see that

$$(*) = \int_{\mathcal{C}} \frac{\partial u^s(x,y)}{\partial\nu(y)}u^c(y)\,\mathrm{d}s(y) + \int_{\partial D_-\setminus\mathcal{C}} u^s(x,y)\frac{\partial u^c}{\partial\nu}(y)\bigg|^- \mathrm{d}s(y) - \int_{\mathcal{C}} u^s(x,y)\,\frac{\partial u^c}{\partial\nu}(y)\bigg|^- \mathrm{d}s(y)$$

$$= -\int_{\partial D_-} \frac{\partial u^s(x,y)}{\partial\nu_-(y)}u^c(y)\,\mathrm{d}s(y) + \int_{\partial D_-} u^s(x,y)\frac{\partial u^c}{\partial\nu_-}(y)\bigg|^- \mathrm{d}s(y) = 0$$

since $\nu_- = -\nu$ on $\mathcal{C}$ and $u^c = 0$ on $\partial D_-\setminus\mathcal{C}$. Hence, we have shown that

$$u^c = \int_{\mathcal{C}} w(\cdot,y)\left[\frac{\partial u^c}{\partial\nu}(y)\right]_{\mathcal{C}} \mathrm{d}s(y) = \mathrm{SL}\left(\left[\frac{\partial u^c}{\partial\nu}(y)\right]_{\mathcal{C}}\right) \qquad \text{in } H^1(D_h^+). \tag{III.27}$$

Consequently, $u^c|_{\mathcal{C}} = g = S\left([\partial u^c/\partial\nu\,(y)]_{\mathcal{C}}\right)$. From Corollary III.4 we obtain the estimate

$$\|g\|_{\widetilde{H}^{1/2}(\mathcal{C})} \leq C\left\|\left[\frac{\partial u^c}{\partial\nu}(y)\right]_{\mathcal{C}}\right\|_{H^{-1/2}(\mathcal{C})} \tag{III.28}$$

which shows, again by Corollary III.4, that the single layer operator is bounded from $H^{-1/2}(\mathcal{C})$ into $\widetilde{H}^{1/2}(\mathcal{C})$. Moreover, we proved that a solution of the problem (III.15) gives rise to a solution of the integral equation (III.23). The opposite direction is shown in the remainder of the proof.

Assume that ϕ solves the integral equation (III.23). Then we know that the single layer operator $\mathrm{SL}\,\phi$ solves the Helmholtz equation in D_h^+, the upwards radiation condition (III.2) and that $\mathrm{SL}\,\phi$ takes the boundary values g on $\mathcal{C}$ and vanishes on $\Gamma_c \setminus \mathcal{C}$. Consequently, the single layer potential $\mathrm{SL}\,\phi$ belongs to $H^1_{\mathrm{loc}}(D_h^+)$. For simplicity we set $u^c = \mathrm{SL}\,\phi$. It remains to show that u^c belongs to $H^1(D_h^+)$ and therefore we consider the decay of $w(x,y)$ for $y \in \mathcal{C}$. Since $\mathcal{C}$ is bounded there is according to (III.8) a constant C such that $|w(x,y)|$, $|\nabla_x w(x,y)| \leq C|x_1|^{-3/2}$ for x large enough and $y \in \mathcal{C}$. Therefore u^c decays as $|x_1|^{-3/2}$, too, which implies that $\|u^c\|_{H^1(D_h^+)}$ is bounded. Now, since u^c satisfies the upward radiation condition we know from [14, 6] that $T(u^c|_{\Gamma_h}) = \partial u^c/\partial\nu|_{\Gamma_h}$ and hence u^c satisfies (III.15). □

We note the following jump relation, which follows from (III.27): For $u = \mathrm{SL}\,\phi$, $\phi \in H^{-1/2}(\mathcal{C})$, it holds

$$\left[\frac{\partial u}{\partial \nu}\right]_{\mathcal{C}} = \phi.$$

The missing minus sign on the right hand side of this equation might be a bit inconvenient, but is due to our choice that the normal ν to D_- on the arc $\mathcal{C}$ points downwards, hence into the domain; but $\partial u/\partial\nu|_{\mathcal{C}}^+$ and $\partial u/\partial\nu|_{\mathcal{C}}^-$ denotes the trace taken from the outside and inside of D_-, respectively. Therefore the sign of the jump relation differs from, e.g., [63], and also from the analogous jump relation in Chapter II.

For the Factorization method we require later on the following theorem.

THEOREM III.6. *The single layer operator S has a decomposition $S = S_0 + K_0$ with K_0 compact and S_0 bounded and coercive: there exists $c > 0$ such that* $\mathrm{Re}\,\langle S_0\phi, \phi\rangle \geq c\|\phi\|^2_{H^{-1/2}(\mathcal{C})}$ *for all $\phi \in H^{-1/2}(\mathcal{C})$. Hence, S is a Fredholm operator with index 0.*

There are a couple of possibilities to prove Theorem III.6. One can for instance adapt the proof of the same assertion for the single layer potential of the free space Green's function, which is provided in [63, Theorem 7.6]. Since, for our situation, a proof can be given completely analogous to that one in [63, Theorem 7.6], we omit this proof. As another option one can use the result below which shows that for small positive wavenumbers S itself is coercive, and note that the difference of two single layer potentials for different positive wavenumbers is compact.

If the wavenumber $k > 0$ is small enough, then S is even coercive. This relies on a Poincaré inequality, see [14, Theorem 3.4] or in a more general context [68], which is considered in the next Chapter IV in more detail.

LEMMA III.7. *For $h > f_+$ we set $H_0^1(D_h^+) := \{u \in H^1(D_h^+) : u|_\Gamma = 0\}$ to be the closed subspace in $H^1(D_h^+)$ with vanishing trace on Γ. On $H_0^1(D_h^+)$ the norm of $H^1(D_h^+)$ is equivalent to the Dirichlet norm*

$$\|u\|_{\mathrm{D}}^2 := \int_{D_h^+} |\nabla u|^2 \, \mathrm{d}x.$$

The latter equivalence implies the existence of a constant C_{D} such that

$$\|u\|^2_{H^1(D_h^+)} \le C_{\mathrm{D}} \|\nabla u\|^2_{L^2(D_h^+)} \qquad \text{for } u \in H_0^1(D_h^+).$$

Due to the representation of T in (III.18) and Plancherel's identity,

$$\begin{aligned} \operatorname{Im}\left(\int_{\Gamma_h} \overline{\phi} T(\phi) \, \mathrm{d}s \right) &= \operatorname{Im}\left(\mathrm{i} \int_{\mathbb{R}} \sqrt{k^2 - \xi^2} \, |\mathcal{F}\phi(\xi)|^2 \, \mathrm{d}\xi \right) \\ &= \int_{|\xi|<k} \sqrt{k^2 - \xi^2} \, |\mathcal{F}\phi(\xi)|^2 \, \mathrm{d}\xi \ge 0, \end{aligned}$$

and therefore $\operatorname{Im} \int_{\Gamma_h} \overline{v} \, \partial v / \partial \nu \, \mathrm{d}s = \operatorname{Im} \int_{\Gamma_h} \overline{v} \, T(v) \, \mathrm{d}s \ge 0$ for any upward radiating function v. The same technique shows that $\operatorname{Re} \int_{\Gamma_h} \overline{v} \, \partial v / \partial \nu \, \mathrm{d}s = \operatorname{Re} \int_{\Gamma_h} \overline{v} \, T(v) \, \mathrm{d}s \le 0$ for any upward radiating function and hence especially for $v = \mathrm{SL}\, \phi$. Using Green's first identity for $v = \mathrm{SL}\, \phi$, we find

$$\langle S\phi, \phi \rangle = \left\langle v, \left[\frac{\partial v}{\partial \nu} \right]_{\mathcal{C}} \right\rangle = \int_{D_h^+ \cup D_-} \left(|\nabla v|^2 - k^2 |v|^2 \right) \mathrm{d}x - \int_{\Gamma_h} v \frac{\partial \overline{v}}{\partial \nu} \, \mathrm{d}s$$

we find

$$\begin{aligned} \operatorname{Re} \langle S\phi, \phi \rangle &\ge \|\nabla v\|^2_{L^2(D_h^+)} - k^2 \|v\|^2_{L^2(D_h^+)} \ge C_{\mathrm{D}}^{-1} \|v\|^2_{H^1(D_h^+)} - k^2 \|v\|^2_{L^2(D_h^+)} \\ &\ge \left(C_{\mathrm{D}}^{-1} - k^2 \right) \|v\|^2_{H^1(D_h^+)} \ge C (C_{\mathrm{D}}^{-1} - k^2) \|\phi\|^2_{H^{1/2}(\mathcal{C})}. \end{aligned} \tag{III.29}$$

Therefore S is coercive for small wavenumbers. Additionally, for arbitrary $k > 0$, we observe that

$$\operatorname{Im} \langle S\phi, \phi \rangle = -\operatorname{Im} \langle v, T(v) \rangle = \operatorname{Im} \langle T(v), v \rangle \ge 0$$

which yields that $\operatorname{Im} S$ is positive semidefinite.

Corollary III.8. *For arbitrary $k > 0$, $\operatorname{Im} S$ is positive semidefinite. Moreover, under the assumption that k^2 is not a Dirichlet eigenvalue of $-\Delta$ in D_-, the single layer operator S is a bijective bounded linear operator between $H^{-1/2}(\mathcal{C})$ and $\widetilde{H}^{1/2}(\mathcal{C})$.*

Proof. Due to Theorem III.6, it remains to show that S is injective for arbitrary wavenumbers. Assume that $\phi \in H^{-1/2}(\mathcal{C})$ is such that $S\phi = 0$ and set $u = \mathrm{SL}\,\phi$. Then $u|_+ = u|_- = 0$ on $\mathcal{C}$. Since u solves the Helmholtz equation in the bounded domain D_-, subject to the Dirichlet condition $u = 0$ on ∂D_- and k^2 is not a Dirichlet eigenvalue in D_- we have $u = 0$ in D_-. Theorem 4.1 in [14] on variational formulations of scattering from a rough surface implies $u = 0$ in $D_+ \setminus \overline{D_c}$, since D_+ satisfies the secondary geometric assumption (III.13). For u vanishing entirely in D, the jump relations for the single layer potential SL imply $\phi = 0$. □

III-4. The Inverse Contamination Problem and a Factorization Method

We are now ready to consider the inverse problem of determing the contamination $\mathcal{C}$ from measured scattered fields. Therefore we introduce a measurement line M above the surface Γ_c, where incident waves in form of point sources are emitted and the resulting scattered waves are measured. The segment M is assumed to be of the form $M = \{(x_1, h) : a < x_1 < b\} \subset \mathbb{R}^2$ for $a < b$ and $h > f_+$. We assume moreover that we know the Green's function $w(x, y)$ of the unperturbed rough surface, especially, we know the surface Γ. Our aim is to apply incident point sources $w(\cdot, y)$ and to reconstruct $\mathcal{C}$ from measurements of the corresponding scattered fields. In view of the application of the Factorization method later on it is well known [48] that we need to modify these incident fields. Indeed, to be able to construct a suitable factorization of the data operator we rely for the moment on the incident fields

$$\overline{w}(\cdot, y) := \overline{G(\cdot, y)} + \overline{u^s(\cdot, y)},$$

which are scattered at Γ_c, resulting in scattered fields $\widetilde{u}^c(\cdot, y)$. The scattered fields $\widetilde{u}^c(\cdot, y)$ solve (III.15) for the data $-\overline{w}(\cdot, y)$.

The use of complex conjugated point sources instead of w seems currently to be necessary for the application of the Factorization method. This is a serious drawback since such sources have no physical relevance. In Section III-5 we are going to link the scattered fields $\widetilde{u}^c(\cdot, y)$ with physically meaningfull fields. For the moment we only consider the (physically irrelevant) near field operator

$$\widetilde{N} : L^2(M) \to L^2(M), \qquad \phi \mapsto \int_M \widetilde{u}^c(\cdot, y)\phi(y)\,\mathrm{d}s(y), \tag{III.30}$$

and pose the *inverse problem* as follows: Given the near field operator $\widetilde{N}$, find the contamination $\mathcal{C}$! The literature on uniqueness results for such inverse contamination problems is rather poor, apart from the studies of obstacle detection in a half space, see, e.g., [31, 56] and the references therein.

Let us first note that the integral operator in (III.30) is compact. Indeed, from Lemma III.1 we know that $w(x, y) = w(y, x)$ for $x, y \in D$ and the arguments of Lemma III.1 also show that $\overline{w}(x, y) + \widetilde{u}^c(x, y) = \overline{w}(y, x) + \widetilde{u}^c(y, x)$. Hence follows the reciprocity relation $\widetilde{u}^c(x, y) = \widetilde{u}^c(x, y)$ for the contamination field. Due to (real) analyticity of any twice continuously differentiable solution of the Helmholtz equation, we observe that the kernel of $\widetilde{N}$ is symmetric and smooth in both variables. Hence, $\widetilde{N}$ is compact on $L^2(M)$.

The derivation of the factorization of $\widetilde{N}$ relies on the data-to-pattern operator L, which maps $\phi \in \widetilde{H}^{1/2}(\mathcal{C})$ to the solution of (III.15) restricted to the measurement line M,

$$L : \widetilde{H}^{1/2}(\mathcal{C}) \to L^2(M), \qquad g \mapsto u^c|_M\,.$$

The auxiliary operator

$$\widetilde{H} : L^2(M) \to \widetilde{H}^{1/2}(\mathcal{C}), \qquad \phi \mapsto \int_M \overline{w}(\cdot, y)\phi(y)\,\mathrm{d}s(y),$$

which corresponds to the Herglotz operator from Section I-3, is easily seen to satisfy $\widetilde{N}\phi = -L\widetilde{H}\phi$. We consider its adjoint

$$\widetilde{H}^* : H^{-1/2}(\mathcal{C}) \to L^2(M), \qquad g \mapsto \int_{\mathcal{C}} w(\cdot, y)g(y)\,\mathrm{d}s(y).$$

$\widetilde{H}^*g$ solves the perturbed boundary value problem (III.15) and takes the boundary values

$$Sg = \int_{\mathcal{C}} w(\cdot, y)g(y)\,\mathrm{d}s(y) \qquad \text{in } \widetilde{H}^{1/2}(\mathcal{C}),$$

which implies $\widetilde{H}^* = LS$ and $\widetilde{N} = -LS^*L^*$. Due to the decompositions $L = \widetilde{H}^*S^{-1}$ and $L^* = S^{-1*}\widetilde{H}$ we can also factorize

$$\widetilde{N} = -\widetilde{H}^*S^{-1}\widetilde{H}, \tag{III.31}$$

whenever S is invertible. Indeed, we consider this factorization in the sequel, since the Herglotz wave function-like operator $\widetilde{H}$ will be of special use in Section III-5.

LEMMA III.9. *Assume that k^2 is not a Dirichlet eigenvalue of $-\Delta$ in D_-. Then $\widetilde{H} : L^2(M) \to \widetilde{H}^{1/2}(\mathcal{C})$ is compact and injective with dense range.*

Proof. The compactness of $\widetilde{H}$ follows from analyticity of the function

$$x \mapsto \int_M \overline{w}(x, y)\phi(y)\,\mathrm{d}s(y)$$

in $D \setminus M$. Injectivity follows from the assumption that k^2 is not a Dirichlet eigenvalue of $-\Delta$ in D_-: Assume that $\widetilde{H}\phi = 0$, that is, $u := \int_M \overline{w}(\cdot, y)\phi(y)\,\mathrm{d}s(y)$ vanishes on $\mathcal{C}$. Since $w(x, \cdot)$ vanishes on Γ we conclude that u vanishes on ∂D_-. In consequence, u satisfies the Helmholtz equation in D_- subject to homogeneous Dirichlet boundary conditions. The assumption that k^2 is not a Dirichlet eigenvalue of D_- implies that $u = 0$ in D_-. On the other hand, the kernel $w(x, y)$ is real analytic for $x \neq y$ and therefore u is real analytic in $D \setminus M$. Because $u = 0$ in the open set D_- it holds $u = 0$ in $\mathbb{R}^2 \setminus M$. Using the jump relations of the single layer potential we conclude that $\phi = [\partial u/\partial \nu]_M = 0$.

To show denseness of the range of $\widetilde{H}$ it is sufficient to show that $\widetilde{H}^* = LS$ is injective. We already know that S is invertible and can restrict ourselves to show that L is injective. Let $\psi \in \widetilde{H}^{1/2}(\mathcal{C})$ be such that $L\psi = u^c|_M = 0$. Since solutions of the Helmholtz equation are real analytic, we infer that v vanishes on the entire horizontal line Γ_h and satisfies the upward radiation condition (III.2). Lemma 2.9 in [16] yields $v = 0$ in $\{x_2 > h\}$. Again, analytic continuation shows $\psi = v|_{\mathcal{C}} = 0$. □

Observe that the decomposition $S = S_0 + K_0$ into a coercive and a compact operator implies that $\langle S^{-1}\psi, \psi\rangle = \langle \phi, S\phi\rangle = \langle \phi, S_0\phi\rangle + \langle \phi, K_0\phi\rangle$ for $\psi \in \widetilde{H}^{1/2}(\mathcal{C})$ and $S\phi = \psi$. Hence,

$$\langle (S^{-1} - S^{-1*}K_0^*S^{-1})\psi, \psi\rangle = \langle \phi, S_0\phi\rangle \geq C\|\phi\|^2_{H^{-1/2}(\mathcal{C})} \geq C\|\psi\|^2_{\widetilde{H}^{1/2}(\mathcal{C})}. \tag{III.32}$$

Therefore S^{-1} is a compact perturbation of a coercive operator, say, $S^{-1} = S_1 + K_1$ with S_1 coercive and K_1 compact. In view of Theorem I.7 on range identities for factorizations we need to consider real and imaginary part of S^{-1} defined by

$$\operatorname{Re} S^{-1} = \frac{1}{2}\left(S^{-1} + S^{-1*}\right), \qquad \operatorname{Im} S^{-1} = \frac{1}{2i}\left(S^{-1} - S^{-1*}\right).$$

Note that $\operatorname{Re}\widetilde{N}$ and $\operatorname{Im}\widetilde{N}$ have factorizations $\operatorname{Re}\widetilde{N} = -\widetilde{H}^*\operatorname{Re} S^{-1}\widetilde{H}$ and $\operatorname{Im}\widetilde{N} = -\widetilde{H}^*\operatorname{Im} S^{-1}\widetilde{H}$, respectively.

THEOREM III.10. *Assume that k^2 is not a Dirichlet eigenvalue of $-\Delta$ in D_-. The near field operator $\widetilde{N}$ has a factorization*

$$\widetilde{N} = -\widetilde{H}^*S^{-1}\widetilde{H}, \tag{III.33}$$

with $\widetilde{H} : L^2(M) \to H^{1/2}(\mathcal{C})$ compact and injective and $S : H^{-1/2}(\mathcal{C}) \to H^{1/2}(\mathcal{C})$. Additionally, $\operatorname{Re} S^{-1}$ is a compact perturbation of a coercive operator and $\operatorname{Im} S^{-1}$ is negative semidefinite. Let $(\lambda_j, \phi_j)_{j\in\mathbb{N}}$ be an eigensystem of the positive, compact and selfadjoint operator

$$\widetilde{N}_\sharp : L^2(M) \to L^2(M), \qquad \widetilde{N}_\sharp = |\operatorname{Re}\widetilde{N}| + \operatorname{Im}\widetilde{N}.$$

Then a point $y \in D$ belongs to D_- if and only if

$$\sum_{j=1}^{\infty} \frac{\left|\langle w(\cdot, y), \phi_j\rangle_{L^2(M)}\right|^2}{\lambda_j} < \infty. \tag{III.34}$$

Of course, the proof of the Factorization method relies again on the abstract Theorem I.7 on range identities in factorizations.

Proof. Since the middle operator S^{-1} maps from $H^{1/2}(\mathcal{C})$ to $H^{-1/2}(\mathcal{C})$, we do not need to lift the factorization into $L^2(\mathcal{C})$ as in the proof of

Theorem II.19. The remark after Lemma III.7 states that $\operatorname{Im} S$ is positive semidefinite. Since

$$\operatorname{Im} \langle S^{-1}\psi, \psi\rangle = -\operatorname{Im} \langle S\phi, \phi\rangle \leq 0 \qquad \text{(III.35)}$$

for $\psi = S\phi$, $\operatorname{Im} S^{-1}$ is negative semidefinite. We already showed in (III.32) that S^{-1} is sum of a coercive and a compact operator. Finally, we recall from Corollary III.8 that S^{-1} is injective, since k^2 is assumed not to be a Dirichlet eigenvalue of $-\Delta$ in D_-.

By the properties of $\widetilde{H}$ and S^{-1}, Theorem I.7 implies that

$$\operatorname{Range}(\widetilde{N}_\sharp)^{1/2} = \operatorname{Range} \widetilde{H}^*.$$

However, the range of $\widetilde{H}^*$ is linked with the contamination $\mathcal{C}$. Indeed, it is easy to see that $\operatorname{Range} \widetilde{H}^* = \operatorname{Range} L$ since $\widetilde{H}^* = LS$ and S is an isomorphism. Consider now a point source $w(\cdot, y)$ at $y \in D$. For $y \in D_+$, $w(\cdot, y)|_M$ cannot belong to $\operatorname{Range} L$: If $Lg = w(\cdot, y)|_M$, analytic continuation together with the upward radiation condition implies that the Herglotz wave function $\int_{\mathcal{C}} w(\cdot, y)\phi(y)\, \mathrm{d}s$, for ϕ such that $S\phi = g$, has a singularity in y. For $y \in D_-$ the singularity of $w(\cdot, y)$ is outside of D_+ and $\widetilde{H}S^{-1}\, w(\cdot, y)|_{\mathcal{C}} = w(\cdot, y)|_M$. Using Picard's range criterion we finally obtain the claimed characterization of D_- and $\mathcal{C}$ in (III.34). □

III-5. Approximation of the Sources

The near field operator $\widetilde{N}$ from (III.30) is physically meaningless because it relies on the complex conjugate of the upward radiating Green's function $w(x, y)$. However, for the physically meaningfull near field operator

$$N : L^2(M) \to L^2(M), \quad \phi \mapsto \int_M u^c(\cdot, y)\phi(y)\, \mathrm{d}s(y),$$

where $u^c(\cdot, y)$ is the scattered field corresponding to the incident point source $w(\cdot, z)$, the Factorization method is not directly applicable [71], merely for algebraic reasons: The factorization $N = -H^t S^{-1} H$, which is analogous to the one of $\widetilde{N}$, involves the transpose H^t rather than the adjoint H^* of

$$H : L^2(M) \to \widetilde{H}^{1/2}(\mathcal{C}), \qquad \phi \mapsto \int_M w(\cdot, y)\phi(y)\,\mathrm{d}s(y).$$

Therefore Theorem I.7 is not applicable. This annoyance can be tackled with a regularization technique that we learned from [48], where the idea is, roughly speaking, to approximate $\widetilde{H}$ by H. For this approximation procedure we choose a test domain Ω such that $M \cap \Omega = \emptyset$ and assume that we have the a-priori information $D_- \subset \Omega$. Moreover, $\Omega \cap D$ is assumed to be a Lipschitz domain such that k^2 is not a Dirichlet eigenvalue of $-\Delta$ in $\Omega \cap D$. By Σ we denote the part of the boundary of Ω contained in D: $\Sigma := \partial\Omega \cap D$, compare Figure III.4.

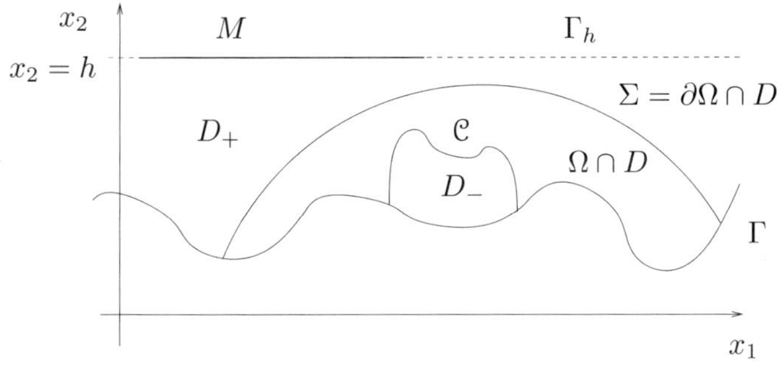

Figure III.4: The test domain $\Omega \cap D$ with boundary $\Sigma = \partial\Omega \cap D$.

Next we define two compact operators V and $\widetilde{V} : L^2(M) \to \widetilde{H}^{1/2}(\Sigma)$,

$$V\phi = \int_M w(\cdot, y)\phi(y)\,\mathrm{d}s, \qquad \widetilde{V}\phi = \int_M \overline{w}(\cdot, y)\phi(y)\,\mathrm{d}s.$$

These two operators evaluate single layer potentials defined on M on the boundary Σ of the test domain,

$$(\mathrm{SL}\,\phi)(x) = \int_M w(x, y)\phi(y)\,\mathrm{d}s, \quad (\widetilde{\mathrm{SL}}\phi)(x) = \int_M \overline{w}(x, y)\phi(y)\,\mathrm{d}s,$$

for $x \notin M$.

Lemma III.11. *V and $\widetilde{V}$ are compact and injective and their ranges are dense in $\widetilde{H}^{1/2}(\Sigma)$.*

We do not give a proof of this result, since the arguments are literally the same as in the proof of Lemma III.9. Even if the ranges of V and $\widetilde{V}$ are both dense in $\widetilde{H}^{1/2}(\Sigma)$, one can in general not solve the equation $V\psi = \widetilde{V}\phi$ for ψ. A way out is to consider a regularized solution of the latter equation, which is done here via Tikhonov regularization, any other regularization via filter functions being applicable, too. More precisely, let

$$P_\delta\phi = (\delta + V^*V)^{-1}V^*\widetilde{V}\phi, \qquad \delta > 0,$$

be the Tikhonov regularization of the equation $V\psi = \widetilde{V}\phi$. From standard regularization theory we know that $P_\delta\phi$ does in general not converge as $\delta \to 0$. However, $VP_\delta\phi$ converges to $\widetilde{V}\phi$ in $\widetilde{H}^{1/2}(\Sigma)$ for all $\phi \in L^2(M)$. Consequently, the boundary values of the single layer potential $\mathrm{SL}\,P_\delta\phi$ on Σ converge to those of $\widetilde{\mathrm{SL}}\phi$. Since both $\mathrm{SL}\,P_\delta\phi$ and $\widetilde{\mathrm{SL}}\phi$ solve the Helmholtz equation in $\Omega\cap D$ and k^2 is not a Dirichlet eigenvalue in $\Omega\cap D$, continuous dependence of the solution from its boundary values implies that also $HP_\delta\phi \to \widetilde{H}\phi$ as $\delta \to 0$. Thus, for every $\phi \in L^2(M)$,

$$\begin{aligned}\lim_{\delta\to 0} NP_\delta\phi &= -\lim_{\delta\to 0} H^{\mathrm{t}}S^{-1}HP_\delta\phi \\ &= -H^{\mathrm{t}}S^{-1}\lim_{\delta\to 0} HP_\delta\phi = -H^{\mathrm{t}}S^{-1}\widetilde{H}\phi = \widetilde{N}\phi.\end{aligned} \tag{III.36}$$

From the point of view of applications it is attractive that $\widetilde{N}$ becomes now in principle available, since N can be measured and P_δ can be computed for arbitrary $\delta > 0$ if a test domain Ω is known a-priori. However, it remains to investigate whether and how the numerically attractive Picard criterion (III.34) is of use when we only know a pointwise approximation NP_δ of $\widetilde{N}$. This question has already been posed in [48]. The pointwise convergence $NP_\delta \to \widetilde{N}$ itself does not directly imply pointwise convergence of $(NP_\delta)_\sharp \to \widetilde{N}_\sharp$, but $NP_\delta\phi \to \widetilde{N}\phi$ implies the weak convergence $P_\delta^*N^*\phi \rightharpoonup \widetilde{N}^*\phi$ for $\phi \in L^2(M)$. However, one can substantially strengthen the convergence properties of NP_δ.

Proposition III.12. *NP_δ converges to $\widetilde{N}$ in norm, that is, $\|NP_\delta - \widetilde{N}\|_{L^2(M)} \to 0$ as $\delta \to 0$.*

Proof. First, we recall that $NP_\delta\phi \to \widetilde{N}\phi$ for all $\phi \in L^2(M)$ yields

$$\langle \phi, P_\delta^* N^* \psi\rangle = \langle NP_\delta \phi, \psi\rangle \to \langle \widetilde{N}\phi, \psi\rangle = \langle \phi, \widetilde{N}^* \psi\rangle, \qquad \delta \to 0,$$

and hence weak convergence $P_\delta^* N^* \psi \rightharpoonup \widetilde{N}^* \psi$ for $\psi \in L^2(M)$. Since $(\delta + V^*V)^{-1}$ is selfadjoint, the representation $P_\delta^* = \widetilde{V}^* V(\delta + V^*V)^{-1}$ holds. Using a singular system of the compact operator V, one easily computes that $V(\delta + V^*V)^{-1} = (\delta + VV^*)^{-1}V$. Using the factorization $N^* = -H^* S^{-1*} \widetilde{H}$ we obtain

$$P_\delta^* N^* = -\widetilde{V}^*(\delta + VV^*)^{-1} V H^* S^{-1*} \widetilde{H}. \tag{III.37}$$

The following observation will be the main tool to finish this proof: The operator $H^* : H^{-1/2}(\mathcal{C}) \to L^2(M)$ can be factorized as $H^* = V^* S_\Sigma^{-1*} H_\Sigma^*$, or equivalently, $H = H_\Sigma S_\Sigma^{-1} V$. Here, S_Σ is the single layer operator on the boundary Σ of the test domain $\Omega \cap D$,

$$S_\Sigma : H^{-1/2}(\Sigma) \to \widetilde{H}^{1/2}(\Sigma), \qquad S_\Sigma \phi = \int_\Sigma w(\cdot, y)\phi(y)\, \mathrm{d}s(y).$$

Our assumption on Ω that k^2 is not an eigenvalue of $-\Delta$ in $\Omega \cap D$ by the way implies that S_Σ is indeed invertible. Moreover, H_Σ denotes the evaluation on $\mathcal{C}$ of the single layer potential SL_Σ on Σ,

$$H_\Sigma : H^{-1/2}(\Sigma) \to \widetilde{H}^{1/2}(\mathcal{C}), \qquad H_\Sigma \phi = \int_\Sigma w(\cdot, y)\phi(y)\, \mathrm{d}s(y).$$

To show that the above equality $H = H_\Sigma S_\Sigma^{-1} V$ holds, we choose $\phi \in L^2(M)$ and set $\psi = V\phi$. As $\mathrm{SL}_\Sigma\, S_\Sigma^{-1}\psi\big|_\Sigma = \psi$ we find that the upwards radiating function $\mathrm{SL}_\Sigma\, S_\Sigma^{-1}\psi$ equals $V\phi$ on Σ and 0 on Γ. The potential $\mathrm{SL}\,\phi$ satisfies the same boundary conditions on Σ and Γ. Exploiting the fact that k^2 is not a Dirichlet eigenvalue in $\Omega \cap D$, we find that both potentials are equal in $\Omega \cap D$. Thus, their evaluations on $\mathcal{C}$ equal, too. This implies the equality $H = H_\Sigma S_\Sigma^{-1} V$.

Using the factorization of H^* together with (III.37) we get

$$P_\delta^* N^* = -\widetilde{V}^*(\delta + VV^*)^{-1} VV^* S_\Sigma^{-1*} H_\Sigma^* S^{-1*} \widetilde{H}.$$

Now, using for instance a singular system of V, one observes that $(\delta + VV^*)^{-1}VV^*$ converges pointwise to the identity. Since $S_\Sigma^{-1*} H_\Sigma^* S^{-1*} \widetilde{H}$

is compact, Theorem 10.6 in [52] implies that $P_\delta^* N^*$ converges in norm. As $P_\delta^* N^* \phi \rightharpoonup \widetilde{N}^* \phi$ and weak and strong limit coincide, the only possible limit of $P_\delta^* N^*$ is $\widetilde{N}^*$. Exploiting that $\|NP_\delta - \widetilde{N}\| = \|P_\delta^* N^* - \widetilde{N}^*\|$ we obtain the claim of the proposition. □

By the norm convergence $NP_\delta \to \widetilde{N}$ and (I.41) it is now clear from (I.41) that also $(NP_\delta)_\sharp \to \widetilde{N}_\sharp$ in norm. Hence we can apply the perturbation theory for the Factorization method developed in Chapter I. Recall that Theorem I.12 shows that one can apply a regularized series criterion for the perturbed data $(NP_\delta)_\sharp$ to reconstruct the perturbation $\mathcal{C}$ asymptotically as $\delta \to 0$. Let us in view of this aim denote by $(\lambda_j^\delta, \psi_j^\delta)_{j\in\mathbb{N}}$ a singular system of $(NP_\delta)_\sharp : L^2(M) \to L^2(M)$. We denote the approximation error by $d_n := \|(NP_\delta)_\sharp - \widetilde{N}_\sharp\|$ and define the truncation index for the regularization Picard criterion by T_n. This index is chosen accordingly to (I.43) and the auxiliary parameters in (I.43) are set as in the corresponding Section I-6, replacing of course $F_\sharp^n$ and $F_\sharp$ by NP_δ and $\widetilde{N}$, respectively. Theorem I.12 yields the following result.

THEOREM III.13. *Denote by $(\delta_n)_{n\in\mathbb{N}}$ a positive zero sequence and by $(\lambda_j^{\delta_n}, \psi_j^{\delta_n})_{j\in\mathbb{N}}$ a singular system of $(NP_{\delta_n})_\sharp$. Choose the truncation index T_n according to* (I.43) *(compare the discussion above) and $y \in \mathbb{R}^2$. Then the sequence*

$$n \mapsto \sum_{j=1}^{T_n} \frac{\left|\langle w(\cdot, y), \psi_j^{\delta_n}\rangle_{L^2(M)}\right|^2}{\lambda_j^{\delta_n}} \quad \textit{is bounded if and only if } y \in D_-.$$

CHAPTER IV

The Factorization Method in Inverse Rough Surface Scattering

IV-1. Introduction

In the previous chapter, the Factorization method was set up to solve an inverse scattering problem for a locally contaminated rough surface. In this chapter we show that the Factorization method can also be used to entirely determine a rough surface from near field measurements. One might argue whether this approach makes sense in practice, since the determination of an infinite structure from near field data seems to be even more ill-posed than the inverse scattering problem for a bounded obstacle, and it is surely hopeless to obtain good reconstructions far away from the measurement points. On the other hand, the material provided in this chapter shows the potential of the Factorization method to provide an estimate of the rough surface in a certain neighborhood of the measurement points, even if no a-priori knowledge on the surface is available. However, note that we are only able to prove the characterization of the rough surface via the Factorization method in case that the rough surface is not too far away from a straight line, where the precise distance is depending on the frequency. A frequency independent characterization is only possible for surfaces which decay at infinity.

To give a brief outlook on this chapter, first we are going to study boundedness of the single layer operator in Sobolev spaces on a rough surface.

Afterwards we show coercivity of the single layer potential between certain Sobolev spaces, a result which is important for the construction of the Factorization method tackled next. One of the important techniques here are again variational formulations for rough surface scattering problems, already introduced in the last section. The inverse problem of determining the rough surface from measurements of scattered fields a finite distance away from the structure is announced on page 124. For a rough surface which is graph of a Lipschitz continuous function and wavenumbers such that the single layer potential is coercive, we show a Factorization method in Section IV-4. For rough surfaces which look like a straight line at infinity, we construct a wavenumber independent Factorization method. The reason for these two restrictive conditions, either for the wavenumber or for the geometry of the surface, is a certain lack of compactness due to the unboundedness of the domain. Finally, we discuss in Section IV-5 why an approach in weighted spaces does not seem to be able to overcome this difficulty.

IV-2. Mapping Properties of the Single Layer Potential

As in the previous chapters, we work in a two dimensional setting and denote by

$$\Gamma := \{(x_1, f(x_1)) : x_1 \in \mathbb{R}\} \subset \mathbb{R}^2$$

the rough surface which is given as the graph of a function $f : \mathbb{R} \to \mathbb{R}$. We choose $f \in C^{0,1}(\mathbb{R}) := \{\phi \in L^\infty(\mathbb{R}), \|\phi'\|_\infty < \infty\}$ to be a Lipschitz continuous function and assume that there are constants $f_+, f_- > 0$ such that $f_- \leq f(x) \leq f_+$ for $x \in \mathbb{R}$. As in the previous chapter, let $\Gamma_h := \{(x_1, x_2) : x_2 = h\} \subset \mathbb{R}^2$ and $U_h = \{(x_1, x_2) : x_2 > h\}$, as well as

$$D := \{(x_1, x_2) : x_2 > f(x_1)\} \subset \mathbb{R}^2$$

and

$$D_h := D \setminus \overline{U_h} = \{(x_1, x_2) : f(x_1) < x_2 < h\} \qquad \text{for } h > f_+,$$

$$D_- := U_0 \setminus \overline{D} = \{(x_1, x_2) : 0 < x_2 < f(x_1)\}.$$

The unit normal normal field ν points downwards on Γ and upwards on Γ_h and is hence the exterior unit normal field to D_h, compare Figure IV.1.

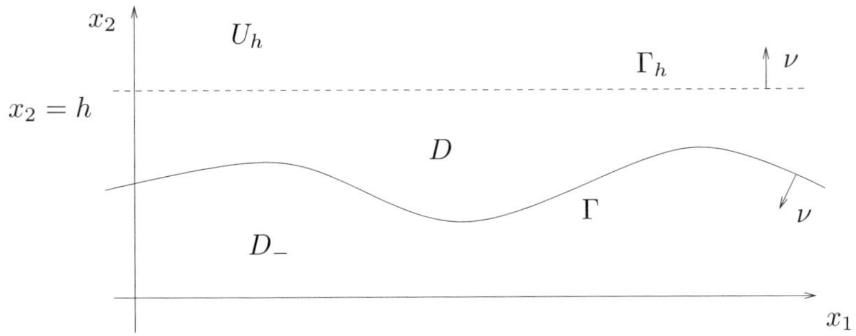

Figure IV.1: The rough surface Γ, which defines the domain D. The domain between the x_1-axis and Γ is called D_-. The normal ν points downwards on Γ and upwards on Γ_h, thus, ν is the exterior normal to $D_h = D \setminus \overline{U_h}$.

We recall from the previous chapter the definition of the Green's function for the half space

$$G(x,y) = \frac{\mathrm{i}}{4}\left(H_0^{(1)}(k|x-y|) - H_0^{(1)}(k|x-y'|)\right), \quad x \neq y,$$

where $y' = (y_1, -y_2)$ and the wavenumber $k > 0$ is again positive. Recall from the last section that $G(\cdot, y)$ and $\nabla_x G(\cdot, y)$ decay as $(1+x_1)^{-3/2}$ for $|x_1| \to \infty$. Using

$$P(x) := \frac{e^{\mathrm{i}k|x|}}{\pi} \int_0^\infty \frac{t^{-1/2} e^{-k|x|t}(1 + (1+\mathrm{i}t)x_2/|x|)}{\sqrt{t - 2\mathrm{i}}(t - \mathrm{i}(1 + x_2/|x|))^2}\, \mathrm{d}t, \qquad x_2 \geq 0,$$

we moreover define the *impedance Green's function G_i for the half space,*

$$G_\mathrm{i}(x,y) = G(x,y) + P(x - y'), \qquad x \neq y \in U_0.$$

The impedance Green's function satisfies the impedance boundary condition [13, Section 3]

$$\frac{\partial G_\mathrm{i}}{\partial x_2}(x,y) + \mathrm{i}k\, G_\mathrm{i}(x,y) = 0 \qquad x \in \Gamma_0,\ y \in U_0,$$

and, according to [15, Lemma 3.1], G_i satisfies the same decay bounds as G,

$$|G_\mathrm{i}(x,y)|\,,\ |\nabla_x G_\mathrm{i}(x,y)| \leq C \frac{(1+x_2)(1+y_2)}{1 + |x-y|^{3/2}}, \quad |x-y| > 1. \qquad \text{(IV.1)}$$

It is remarked in [64, Section 3] that any partial derivative of $G(\cdot, y)$ decays with the same order $-3/2$, and the argument given in that reference shows that this also holds for $G_{\mathrm{i}}(\cdot, y)$.

In the following, we frequently use the single layer potential for the Green's function G_{i}, formally defined by

$$(\mathrm{SL}\,\phi)(x) := \int_\Gamma G_{\mathrm{i}}(x,y)\phi(y)\,\mathrm{d}s(y), \qquad x \in D.$$

From [85] we know that for $\phi \in BC(\Gamma)$, $\mathrm{SL}\,\phi$ is a twice continuously differentiable function in D which solves the Helmholtz equation. Moreover, $\mathrm{SL}\,\phi$ can be continuously extended up to the boundary with limiting values

$$(S\phi)(x) := \int_\Gamma G_{\mathrm{i}}(x,y)\phi(y)\,\mathrm{d}s(y), \qquad x \in \Gamma,$$

and as usual we call $S\phi$ the single layer operator with density ϕ. Transforming the integral to an integral on the real axis, we obtain

$$(S\phi)(s, f(s)) = \int_{\mathbb{R}} G_{\mathrm{i}}\big((s,f(s)),(t,f(t))\big)\phi(t,f(t))\sqrt{1+|f'(t)|^2}\,\mathrm{d}t$$

and writing $\hat{\phi}(t) = \phi(t, f(t))$ and $(\hat{S}\hat{\phi})(s) = (S\phi)(s, f(s))$ yields

$$(\hat{S}\hat{\phi})(s) = \int_{\mathbb{R}} \hat{G}_{\mathrm{i}}(s,t)\hat{\phi}(t)\sqrt{1+|f'(t)|^2}\,\mathrm{d}t, \qquad s \in \mathbb{R},$$

with $\hat{G}_{\mathrm{i}}(s,t) := G_{\mathrm{i}}((s,f(s)),(t,f(t)))$. Therefore $\hat{S}$ is the operator of multiplication by $\sqrt{1+|f'|^2}$ composed with an integral operator. Under our assumptions on Γ, multiplication by $\sqrt{1+|f'|^2}$ is a bounded linear operation on $L^2(\mathbb{R})$, since any Lipschitz continuous function $f \in C^{0,1}(\mathbb{R})$ has a bounded measurable weak derivative [25, Theorem 4.5.8]. Moreover, the singularity of G_{i} is of logarithmic type, see, e.g., [22, Equation 3.61], and $G_{\mathrm{i}}(s,t)$ decays as $|s-t|^{-3/2}$ due to (IV.1). Thus we observe that $|\hat{G}_{\mathrm{i}}(s,t)| \le \ell(s-t)$ for some function $\ell \in L^1(\mathbb{R})$ and therefore Young's inequality [42, Theorem IV.1.6] ensures that

$$\hat{\phi} \mapsto \int_{\mathbb{R}} \hat{G}_{\mathrm{i}}(\cdot,t)\hat{\phi}(t)\,\mathrm{d}t$$

is a bounded operation on $L^2(\mathbb{R})$. Consequently, $\hat{S}$ is a bounded operator on $L^2(\mathbb{R})$ and S is bounded on $L^2(\Gamma)$, since by definition

$$\|S\phi\|^2_{L^2(\Gamma)} = \int_{\mathbb{R}} |(S\phi)(t, f(t))|^2 \sqrt{1 + |f'(t)|^2}\,\mathrm{d}t.$$

We also establish that S is a bounded operator from $L^2(\Gamma)$ into $H^1(\Gamma)$. Therefore we first recall that for $s \in [0,1]$, by definition,

$$H^s(\Gamma) = \{\psi \in L^2(\Gamma) : \psi(x_1, f(x_1)) \in H^s(\mathbb{R})\} \tag{IV.2}$$

and for $s \in [-1, 0)$ the space $H^s(\Gamma)$ is defined by duality [63, page 98] using the inner product in $L^2(\Gamma)$. In the arguments leading to mapping properties of S, as well as later on in this chapter, we extensively make use of families of cut-off functions. Let $\chi \in C_c^\infty(\mathbb{R})$ with

$$0 \le \chi \le 1, \qquad \chi(t) = 0 \quad \text{for } |t| > 2, \qquad \chi(t) = 1 \quad \text{for } |t| < 1. \tag{IV.3}$$

For later purposes, we choose χ such that

$$\sum_{n \in 3\mathbb{Z}} \chi_n \equiv 1 \qquad \text{where } \chi_n(\cdot) := \chi_n(\cdot - n),\ n \in \mathbb{Z}.$$

We choose another cut-off function $\chi^1 \in C_c^\infty(\mathbb{R})$ which satisfies

$$0 \le \chi^1 \le 1, \quad \chi^1(t) = 1 \text{ for } t \in \operatorname{supp}(\chi), \quad \chi^1(t) = 0 \text{ for } t \notin 2\operatorname{supp}(\chi),$$

and define $\chi^1_n(\cdot) := \chi^1(\cdot - n)$. The functions χ_1 and χ_2 are sketched in Figure IV.2. Finally, let $\chi^2 \in C_c^\infty(\mathbb{R})$ such that $0 \le \chi^2 \le 1$, $\chi^2 = 1$

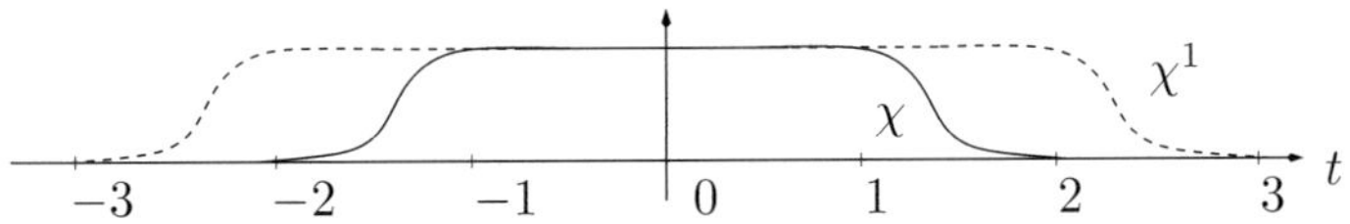

Figure IV.2: The cut-off functions χ and χ^1. Note that $\chi^1(t) = 1$ for $t \in \operatorname{supp}(\chi)$.

on the set $\{t \in \mathbb{R} : \operatorname{dist}(t, \operatorname{supp}(\chi^1)) \le 2 \operatorname{diam} \operatorname{supp}(\chi)\}$. Again, we set

$\chi_n^2 = \chi^2(\cdot - n)$. A technical computation shows that

$$\begin{aligned}\|\hat{\psi}\|^2_{H^1(\mathbb{R})} &= \int_{\mathbb{R}} |\hat{\psi}|^2 + |\hat{\psi}'|^2 \,\mathrm{d}t \\ &= \sum_{n\in 3\mathbb{Z}} \int_{\mathbb{R}} \left(\chi_n|\hat{\psi}|^2 + \chi_n|\hat{\psi}'|^2\right) \mathrm{d}t \\ &\leq \sum_{n\in 3\mathbb{Z}} \int_{\mathbb{R}} \left(|\chi_n^1\hat{\psi}|^2 + |(\chi_n^1\hat{\psi})'|^2\right) \mathrm{d}t \\ &= \sum_{n\in 3\mathbb{Z}} \|\chi_n^1\hat{\psi}\|^2_{H^1(\mathbb{R})} \leq C(\chi^1)\|\hat{\psi}\|^2_{H^1(\mathbb{R})}, \quad \hat{\psi} \in H^1(\mathbb{R}).\end{aligned} \tag{IV.4}$$

The last step follows from the estimate

$$\begin{aligned}\sum_{n\in 3\mathbb{Z}} \|\chi_n^1\hat{\psi}\|^2_{H^1(\mathbb{R})} &\leq 2\|\chi^1\|^2_{C^1(\mathbb{R})} \sum_{n\in 3\mathbb{Z}} \int_{\mathrm{supp}(\chi_n^1)} |\hat{\psi}|^2 \,\mathrm{d}t \\ &\quad + 2\|\chi^1\|^2_{C^0(\mathbb{R})} \sum_{n\in 3\mathbb{Z}} \int_{\mathrm{supp}(\chi_n^1)} \left(|\hat{\psi}|^2 + |\hat{\psi}'|^2\right) \mathrm{d}t \leq C(\chi^1)\|\hat{\psi}\|^2_{H^1(\mathbb{R})}.\end{aligned}$$

We conclude that we can localize the norm on $H^1(\mathbb{R})$ by cut-off functions. Recall from the theory of layer potentials for closed bounded surfaces [63, Chapter 6] that

$$\|\hat{S}(\chi_n^2\hat{\phi})\|^2_{H^1(K_n)} \leq C(\chi^2)\|\chi_n^2\hat{\phi}\|^2_{L^2(\mathbb{R})}, \tag{IV.5}$$

where the set K_n is defined as

$$K_n := \left\{t \in \mathbb{R} : \mathrm{dist}(t, \mathrm{supp}(\chi_n^2) \leq \mathrm{diam}(\chi^1)\right\}. \tag{IV.6}$$

Now we split the kernel of $\hat{S}$ with the help of the cut-off function χ in a local and a global part,

$$\hat{G}_{\mathrm{i}}(s,t) = \hat{G}_{\mathrm{i}}^l(s,t) + \hat{G}_{\mathrm{i}}^g(s,t) := \chi(s-t)\hat{G}_{\mathrm{i}}(s,t) + (1-\chi(s-t))\hat{G}_{\mathrm{i}}(s,t), \tag{IV.7}$$

and prove boundedness of the associated operators $\hat{S}^l$ and $\hat{S}^g$ from $L^2(\mathbb{R})$ to $H^1(\mathbb{R})$ for both parts separately. The global part is the easier one, since $(1-\chi(s-t))\hat{G}_{\mathrm{i}}(s,t)$ is a smooth function and all of its partial derivatives decay as $|x|^{-3/2}$. Therefore $\partial/(\partial s)\, \hat{G}_{\mathrm{i}}^{g}(s,t) \in L^1(\mathbb{R})$ as the claim follows as above from Young's inequality,

$$\|\hat{S}^g \hat{\phi}\|_{H^1(\mathbb{R})} \leq C\|\hat{\phi}\|_{L^2(\mathbb{R})}.$$

Combining the latter estimate with (IV.5) yields for the local part $\hat{S}^l = \hat{S} - \hat{S}^g$ that

$$\|\hat{S}^l(\chi_n^2 \hat{\phi})\|_{H^1(K_n)} \leq C\|\chi_n^2 \hat{\phi}\|_{L^2(\mathbb{R})}.$$

Thus, we estimate

$$\begin{aligned} \|\hat{S}^l \hat{\phi}\|^2_{H^1(\mathbb{R})} &\leq \sum_{n\in 3\mathbb{Z}} \|\chi_n^1 \hat{S}^l \hat{\phi}\|^2_{H^1(\mathbb{R})} = \sum_{n\in 3\mathbb{Z}} \|\chi_n^1 \hat{S}^l \hat{\phi}\|^2_{H^1(\mathrm{supp}(\chi_n^1))} \\ &= \sum_{n\in 3\mathbb{Z}} \|\chi_n^1 \hat{S}^l (\chi_n^2 \hat{\phi})\|^2_{H^1(\mathrm{supp}(\chi_n^1))} \\ &\leq C(\chi^1) \sum_{n\in 3\mathbb{Z}} \|\hat{S}^l(\chi_n^2 \hat{\phi})\|^2_{H^1(K_n)} \\ &\leq C(\chi^1,\chi^2) \sum_{n\in 3\mathbb{Z}} \|\chi_n^2 \hat{\phi}\|^2_{L^2(\mathbb{R})} \leq C(\chi^1,\chi^2)\|\hat{\phi}\|^2_{L^2(\mathbb{R})}. \end{aligned} \tag{IV.8}$$

Hence, $\hat{S}$ is bounded from $L^2(\mathbb{R})$ to $H^1(\mathbb{R})$. This implies that S is bounded from $L^2(\Gamma)$ to $H^1(\Gamma)$, by definition of the latter space in (IV.2).

Since the kernel of the $L^2(\Gamma)$-adjoint operator S^* of S is the complex conjugate $\overline{G_{\mathrm{i}}(x,y)}$, it follows that S^* is bounded from $L^2(\Gamma)$ to $H^1(\Gamma)$, too. Using duality in the Gelfand triple $H^{-1}(\Gamma) \subset L^2(\Gamma) \subset H^1(\Gamma)$ we obtain that S is also bounded from $H^{-1}(\Gamma)$ to $L^2(\Gamma)$. By interpolation theory for Sobolev spaces [63, Appendix B] we finally conclude that S is a bounded operator from $H^{-1/2}(\Gamma)$ into $H^{1/2}(\Gamma)$.

By a similar technique as for the single layer operator one can prove that the single layer potential SL is bounded from $H^{-1/2}(\Gamma)$ to $H^1(D_{h_0})$, $h_0 > 0$. As above, we split the integral operator in a local and a global part,

$$\begin{aligned} \mathrm{SL}\,\phi = \mathrm{SL}^l\,\phi + \mathrm{SL}^g\,\phi &:= \int_\Gamma \chi(x_1 - y_1) G_{\mathrm{i}}(x-y)\phi(y)\,\mathrm{d}s(y) \\ &\quad + \int_\Gamma (1 - \chi(x_1 - y_1)) G_{\mathrm{i}}(x-y)\phi(y)\,\mathrm{d}s(y). \end{aligned}$$

From standard theory on the single layer potential on bounded surfaces and the localization technique shown above it follows that the local part SL^l is bounded from $H^{-1/2}(\Gamma)$ into $H^1(D_h)$, and even into $H^1(U_0 \setminus U_{h_0})$ for some $h_0 > f_+$. For the global part with smooth kernel $(1 - \chi(x_1 - y_1))G_{\mathrm{i}}(x-y)$ we use again the decay of the Green's function G_{i} and any of its partial derivatives of order $-3/2$, see (IV.1), to observe that the $L^2(\Gamma_h)$-norm of any partial derivative of $\mathrm{SL}^g\,\phi$ on any line Γ_h, $0 \le h \le h_0$, is bounded in terms of the $H^{-1/2}(\Gamma)$-norm of ϕ. This follows by the same arguments we used to prove boundedness of S^g above. Since $[0, h_0]$ is compact, the constant in the corresponding norm estimate is uniformly bounded in $[0, h_0]$. We conclude that

$$\begin{aligned} \|\,\mathrm{SL}^g\,\phi\|^2_{H^1(U_0\setminus U_{h_0})} &= \int_0^{h_0} \int_{\Gamma_h} \left(|u|^2 + |\nabla u|^2\right) \mathrm{d}x_1\,\mathrm{d}x_2 \\ &\le \int_0^{h_0} C(x_2)\|\phi\|_{H^{-1/2}(\Gamma_{x_2})}\,\mathrm{d}x_2 \le C\|\phi\|_{H^{-1/2}(\Gamma)}, \end{aligned}$$

which yields boundedness of the single layer potential between $H^{-1/2}(\Gamma)$ and $H^1(U_0 \setminus U_{h_0})$.

IV-3. Variational Formulations for Rough Surface Scattering

While integral equation formulations for rough surface scattering in the space of bounded continuous functions have been studied in detail during the last decade, variational formulations for surface scattering problems have only been investigated recently. Partly, we already introduced techniques and results which are relevant for us in the last chapter in Section III-3. We apply these results now to show that the single layer potential is coercive in a certain range of wavenumbers.

Let us recall from Section III-3 the variational formulation of the Helmholtz equation $\Delta u + k^2 u = f$ with $f \in L^2(D_h)$ where $u \in H^1_0(D_h) = \{u \in H^1(D_h) : u|_\Gamma = 0\}$ satisfies the Dirichlet boundary condition $u|_\Gamma = 0$ and the transparent boundary condition $\partial u/\partial\nu|_{\Gamma_h} = T(u|_{\Gamma_h})$. Application of the Dirichlet-to-Neumann operator T on the line Γ_h, introduced in (III.18), yields that the solution $u \in H_0^1(D_h)$ satisfies

$$\mathcal{B}(u,v) := \int_{D_h} \nabla u \nabla \overline{v} - k^2 u \overline{v} \, \mathrm{d}x - \int_{\Gamma_h} \overline{v} T(u) \, \mathrm{d}s = - \int_{D_h} f \overline{v} \, \mathrm{d}x \quad \text{(IV.9)}$$

for all $v \in H_0^1(D_h)$. It is shown in [14] that the sesquilinear form $\mathcal{B}$ satisfies an inf-sup condition and thereby one obtains the following variational result on existence and uniqueness of rough surface scattering problems.

THEOREM IV.1 (Theorem 4.1 in [14]). *The variational problem to find $u \in H_0^1(D_h)$ such that*

$$\mathcal{B}(u,v) = f(v) \qquad \text{for all } v \in H_0^1(D_h) = \{v \in H^1(D_h),\ v|_\Gamma = 0\}$$

has a unique solution for all $f \in H^{-1}(D_h)$, which depends continuously on f,

$$\|u\|_{H^1(D_h)} \leq C \|f\|_{H^{-1}(D_h)}.$$

For small k, such that $0 < k(h - f_-) < \sqrt{2}$, $\mathcal{B}$ is coercive on $H_0^1(D_h)$: for $u \in H_0^1(D_h)$ it holds

$$\operatorname{Re} B(u,u) \geq \frac{2 - k^2(h - f_-)^2}{2 + k^2(h - f_-)^2} \left(\|\nabla u\|^2_{L^2(D_h)} + k^2 \|u\|^2_{L^2(D_h)} \right). \quad \text{(IV.10)}$$

We mention that it is basically the Poincaré inequality [14, Lemma 3.4]

$$\|u\|_{L^2(D_h)} \leq \frac{h - f_-}{\sqrt{2}} \|\nabla u\|_{L^2(D_h)}, \quad \text{(IV.11)}$$

which implies for $0 < k < k_0 := \sqrt{2}/(h - f_-)$ the ellipticity stated in (IV.10). Note that the Poincaré inequality also implies that the $L^2(D_h)$-norm of the gradient defines an equivalent norm in $H^1(D_h)$.

As in Chapter (III) we note the following corollary on solvability of the inhomogeneous Dirichlet problem.

COROLLARY IV.2. *There is a constant $C > 0$ such that for each $\psi \in H^{1/2}(\Gamma)$ there is a unique solution $u \in H^1(D_h)$ of the Dirichlet problem*

$$\Delta u + k^2 u = 0 \quad \text{in } D_h, \qquad u|_\Gamma = \psi, \qquad \left. \frac{\partial u}{\partial \nu} \right|_{\Gamma_h} = T(u),$$

which satisfies $\|u\|_{H^1(D)} \leq C \|\psi\|_{H^{1/2}(\Gamma)}$.

Since the proof of Corollary IV.2 is analogous to that of Corollary III.3, we omit it here. Next we show that for $0 < kf_+ < \sqrt{2}$ the single layer operator is coercive.

PROPOSITION IV.3. *The single layer operator* $S : H^{-1/2}(\Gamma) \to H^{1/2}(\Gamma)$ *is coercive for* $0 < kf_+ < \sqrt{2}$.

Proof. Because $0 < kf_+ < \sqrt{2}$, there is $h > f_+$ such that

$$0 < k \max\{h - f_-, f_+\} < \sqrt{2},$$

and we fix this number h in the following. Set $\Omega_R = \{x \in \mathbb{R}^2 : |x_1| < R,\, 0 < x_2 < h\}$ and $I_{\pm R} = \{x \in \mathbb{R}^2 : x_1 = \pm R,\, x_2 \in (0,h)\}$. Moreover, note the following trace and jump relations* for the single layer operator,

$$\mathrm{SL}\,\phi|_\Gamma = S\phi \qquad \text{and} \qquad \left[\frac{\partial}{\partial\nu}\mathrm{SL}\,\phi\right]_\Gamma = \phi, \tag{IV.12}$$

which hold for $\phi \in BC(\Gamma)$, see [85, Lemma A.2] (and can be extended to $H^{-1/2}(\Gamma)$ by density). For $\phi \in C_c^\infty(\Gamma)$ and $u = \mathrm{SL}\,\phi$, Green's first identity implies

$$\begin{aligned}\int_{\Gamma\cap\Omega_R} \overline{\phi} S\phi \,\mathrm{d}s &= \int_{\Gamma\cap\Omega_R} \left[\frac{\partial \overline{u}}{\partial\nu}\right]_\Gamma u \,\mathrm{d}s \\ &= \int_{\Omega_R} \left(|\nabla u|^2 - k^2|u|^2\right) \mathrm{d}x - \int_{\partial\Omega_R} \overline{u}\frac{\partial u}{\partial\nu}\,\mathrm{d}s. \end{aligned}\tag{IV.13}$$

We rewrite the last boundary integral as follows,

$$\begin{aligned}\int_{\partial\Omega_R} \overline{u}\frac{\partial u}{\partial\nu}\,\mathrm{d}s &= \int_{\Gamma_h} \overline{u}\frac{\partial u}{\partial\nu}\,\mathrm{d}s + \int_{I_R\cup I_{-R}} \overline{u}\frac{\partial u}{\partial\nu}\,\mathrm{d}s - \int_{\{|x_1|>R\}\times\{h\}} \overline{u}\frac{\partial u}{\partial\nu}\,\mathrm{d}s \\ &= \int_{\Gamma_h} \overline{u}\frac{\partial u}{\partial\nu}\,\mathrm{d}s - \int_{\partial A_R^+} \overline{u}\frac{\partial u}{\partial\nu}\,\mathrm{d}s - \int_{\partial A_R^-} \overline{u}\frac{\partial u}{\partial\nu}\,\mathrm{d}s,\end{aligned}$$

where $A_R^\pm = \{(x_1,x_2) : x_1 \gtrless \pm R,\, x_2 \in (0,h)\}$. Trace theorems from $H^1(A_R^\pm)$ into $H^{\pm 1/2}(\partial A_R^\pm)$ imply that

$$\left|\int_{\partial A_R^\pm} \overline{u}\frac{\partial u}{\partial\nu}\,\mathrm{d}s\right| \leq \|u\|_{H^{1/2}(\partial A_R^\pm)} \|\partial u/\partial\nu\|_{H^{-1/2}(\partial A_R^\pm)} \leq C\|u\|^2_{H^1(A_R^\pm)}$$

*By $[u]_\Gamma$ we denote the jump of a function u across Γ, that is, $[u]_\Gamma := u|_\Gamma^+ - u|_\Gamma^-$, where $u|_\Gamma^+$ is the trace on Γ taken from D and $u|_\Gamma^-$ is the trace taken from D_-.

and the latter quantity tends to zero as $R \to \infty$. Note that the limit as $R \to \infty$ of the right hand side of (IV.13) exists since S is bounded from $H^{-1/2}(\Gamma)$ into $H^{1/2}(\Gamma)$. Recall that

$$-\mathrm{Re}\left(\int_{\Gamma_h} u\,\frac{\partial\overline{u}}{\partial\nu}\,\mathrm{d}s\right) = -\mathrm{Re}\left(\int_{\Gamma_h} \overline{u}T(u)\,\mathrm{d}s\right)$$
$$= \int_{|\xi|>k} \sqrt{\xi^2-k^2}\,|\mathcal{F}u(\xi,h)|^2\,\mathrm{d}\xi \geq 0$$

is non-negative. Hence, we conclude as in (III.29) that

$$\mathrm{Re}\left(\int_\Gamma \overline{\phi}S\phi\,\mathrm{d}s\right) = \int_{U_0\setminus U_h}\left(|\nabla u|^2 - k^2|u|^2\right)\mathrm{d}x - \mathrm{Re}\left(\int_{\Gamma_h} u\,\frac{\partial\overline{u}}{\partial\nu}\,\mathrm{d}s\right)$$
$$\geq \int_{D_h}\left(|\nabla u|^2 - k^2|u|^2\right)\mathrm{d}x + \int_{D_-}\left(|\nabla u|^2 - k^2|u|^2\right)\mathrm{d}x$$
$$\geq \frac{2-k^2(h-f_-)}{2}\|\nabla u\|^2_{L^2(D_h)} + \frac{2-k^2f_+}{2}\|\nabla u\|^2_{L^2(D_-)}$$
$$\geq C\|u\|^2_{H^1(D_h)} + C\|u\|^2_{H^1(D_-)}. \quad \text{(IV.14)}$$

Here we exploited again the Poincaré inequality (IV.11), first in D_h and then in D_-, where the constant $(h-f_-)/\sqrt{2}$ from (IV.11) has to be replaced once by $f_+/\sqrt{2}$. The (conormal) trace theorem from $H^1(D_h)$ and $H^1(D_-)$ into $H^{-1/2}(\Gamma)$, see [63, Lemma 4.3], implies

$$\mathrm{Re}\left(\int_\Gamma \overline{\phi}S\phi\,\mathrm{d}s\right) \geq C\left(\left\|\frac{\partial u}{\partial\nu}\Big|_\Gamma^+\right\|_{H^{-1/2}(\Gamma)} + \left\|\frac{\partial u}{\partial\nu}\Big|_\Gamma^-\right\|_{H^{-1/2}(\Gamma)}\right)^2$$
$$\geq C\left\|\left[\frac{\partial u}{\partial\nu}\right]_\Gamma\right\|^2_{H^{-1/2}(\Gamma)}.$$

Hence, using again the trace and jump relations from (IV.12), we infer that

$$\mathrm{Re}\left(\int_\Gamma \overline{\phi}\,S\phi\,\mathrm{d}s\right) \geq C\|\phi\|^2_{H^{-1/2}(\Gamma)},$$

for $\phi \in C_c^\infty(\Gamma)$. By density of this set of functions in $H^{-1/2}(\Gamma)$, we conclude that S is coercive for small wavenumbers. □

IV-4. Factorization of the Near Field Operator and Characterization of the Rough Surface

We are now ready to state the factorization of the near field operator and prove an explicit characterization of Γ in terms of this operator. Therefore we rely on the assumption that the positive wavenumber is small enough compared to the surface elevation of the rough surface such that the single layer operator is coercive. The case when the wavenumber is too large for S to be coercive will be discussed in detail in the next section. Note that the factorization in this chapter is again based on complex conjugate point sources $\overline{G_{\mathrm{i}}(\cdot, y)}$, for the same reasons as in the last chapter, see the discussion in the beginning of Section III-5.

The near field operator N is defined on a smooth bounded measurement line $M = \{(x_1, h) : a < x_1 < b\} \subset D$. Let

$$\widetilde{N} : L^2(M) \to L^2(M), \qquad \phi \mapsto \int_M u^s(\cdot, y)\phi(y)\,\mathrm{d}s, \tag{IV.15}$$

where $u^s(\cdot, y)$ is the scattered field due to the point source $\overline{G_{\mathrm{i}}(\cdot, y)}$. The field $u^s(\cdot, y)$ is by Theorem IV.1 well defined since Γ is the graph of a Lipschitz continuous function.

The operator $\widetilde{N}$ provides the data for the *inverse problem* which we investigate in this chapter. It reads as follows: Given the near field operator $\widetilde{N}$, determine the rough surface Γ! Uniqueness for this inverse problem can be shown using Schiffer's argument, see, e.g., [22]. Note that Schiffer's argument works for the near field operator $\widetilde{N}$ as well as for its counterpart which is defined without complex conjugation of G_{i}.

As in the previous chapters we also consider a Herglotz-like operator

$$\widetilde{H} : L^2(M) \to H^{1/2}(\Gamma), \qquad \widetilde{H}\phi = \int_M \overline{G_{\mathrm{i}}(\cdot, y)}\phi(y)\,\mathrm{d}s,$$

and note that $\widetilde{N} = -L\widetilde{H}$ where L is the data-to-pattern operator which maps $\psi \in H^{1/2}(\Gamma)$ to $u|_M \in L^2(M)$, where $u \in H^1(D_h)$ denotes the unique radiating continuation to D of the solution to the rough surface scattering problem

$$\Delta u + k^2 u = 0 \quad \text{in } D_h, \qquad u|_\Gamma = \psi, \qquad T(u) = \left.\frac{\partial u}{\partial \nu}\right|_{\Gamma_h}. \tag{IV.16}$$

As M is not necessarily included in D_h, the restriction $u|_M \in L^2(M)$ eventually requires extension of u to D by the formula given in (III.17). Since this extension is based on upward propagating plane waves, we call it the radiating extension of u to D.

Since $\widetilde{H}^* = LS$ we obtain the factorization

$$\widetilde{N} = -LS^*L^* \tag{IV.17}$$

and whenever k, h and $f_\pm$ are such that Proposition IV.3 holds,

$$\widetilde{N} = -\widetilde{H}^* S^{-1} \widetilde{H}. \tag{IV.18}$$

In this section, we prefer the factorization given in (IV.18) instead of the one in (IV.17). However, (IV.17) does also hold in cases where S is not invertible, therefore we consider (IV.17) in the next section when discussing limitations of the Factorization method.

LEMMA IV.4. *$\widetilde{H}$ is injective and compact from $L^2(M)$ into $H^{-1/2}(\Gamma)$.*

The proof that $\widetilde{H}$ is injective and compact follows the lines of the proof of Lemma III.9, except that we can drop here the assumption that k^2 is not a Dirichlet eigenvalue of D_-, since we use the impedance Green's function G_{i}, compare [85, Appendix C]. By the way, this lemma is the only part in this chapter where we exploit the use of the impedance Green's function instead of the Dirichlet Green's function for the half space. Using this auxiliary result we can now prove the following characterization for Γ.

THEOREM IV.5. *Let $0 < kf_+ < \sqrt{2}$. Then the near field operator $\widetilde{N}$ has a factorization*

$$\widetilde{N} = -\widetilde{H}^* S^{-1} \widetilde{H},$$

with $\widetilde{H} : L^2(M) \to H^{-1/2}(\Gamma)$ compact and injective and

$$S^{-1} : H^{-1/2}(\Gamma) \to H^{1/2}(\Gamma).$$

Additionally, $\operatorname{Re} S^{-1}$ *is a coercive operator and* $\operatorname{Im} S^{-1}$ *is negative semi-definite. Denote by $(\lambda_j, \phi_j)_{j\in\mathbb{N}}$ an eigensystem of the positive, compact and selfadjoint operator*

$$\widetilde{N}_\sharp : H^{-1/2}(M) \to H^{1/2}(M), \qquad \widetilde{N}_\sharp = |\operatorname{Re} \widetilde{N}| + \operatorname{Im} \widetilde{N}.$$

Then a point $y \in U_0$ belongs to D_- if and only if

$$\sum_{j=1}^{\infty} \frac{\left|\langle G_{\mathrm{i}}(\cdot, y), \phi_j \rangle_{L^2(M)}\right|^2}{\lambda_j} < \infty. \tag{IV.19}$$

This characterization in inverse rough surface scattering relies of course again on Theorem I.7. We remark that the claim of the theorem holds also when G_{i} in (IV.19) is replaced by G, the Dirichlet Green's function for the half space. However, one cannot replace G_{i} by the free-space fundamental solution $\Phi(\cdot, y)$, since the trace $\Phi(\cdot, y)|_\Gamma$ does not belong to $H^{1/2}(\Gamma)$.

Proof. The operator S is coercive by Proposition IV.3. This implies that the selfadjoint part $\operatorname{Re} S$ is coercive. The non-selfadjoint part $\operatorname{Im} S$ is negative semidefinite: Using Green's first identity as in (IV.14) and the representation (III.18) for the Dirichlet-to-Neumann operator T,

$$\begin{aligned} \operatorname{Im}\left(\int_\Gamma \overline{\phi} S \phi \,\mathrm{d}s\right) &= \operatorname{Im}\left(\int_{\Gamma_h} \overline{u} T(u) \,\mathrm{d}s\right) \\ &= \operatorname{Im}\left(\mathrm{i} \int_{\mathbb{R}} \sqrt{k^2 - \xi^2}\, |\mathfrak{F}u(\xi, h)|^2 \,\mathrm{d}\xi\right) \\ &= \int_{|\xi|<k} \sqrt{k^2 - \xi^2}\, |\mathfrak{F}u(\xi, h)|^2 \,\mathrm{d}\xi \geq 0, \end{aligned}$$

where $u = \mathrm{SL}\,\phi$. Consequently, $\operatorname{Im} S$ is positive semidefinite. In precisely this situation we showed in the last chapter in (III.32) and (III.35) that $\operatorname{Re} S^{-1}$ is coercive and $\operatorname{Im} S^{-1}$ is negative semidefinite, respectively. Literally the same arguments apply also at this point. In view of the compactness and injectivity of $\widetilde{H}$, see Lemma IV.4, we can apply Theorem I.7 which yields that the range of the square root of

$$\widetilde{N}_\sharp = |\operatorname{Re} N| + \operatorname{Im} N$$

equals the range of $\widetilde{H}^*$.

The proof is hence complete if we can show that $G_{\mathrm{i}}(\cdot, y)|_M$ belongs to the range of $\widetilde{H}^*$ if and only if $y \notin \overline{D}$. This can be seen analogously as in the proof of Theorem III.10. Indeed, for $y \notin \overline{D}$ it holds that $\widetilde{H}^* S^{-1}\, G_{\mathrm{i}}(\cdot, y)|_\Gamma = G_{\mathrm{i}}(\cdot, y)|_M$. Further, if $G_{\mathrm{i}}(\cdot, y)|_M = \widetilde{H}^* \phi$, then analytic continuation and the upward radiation condition imply that $G_{\mathrm{i}}(\cdot, y)$

belongs to $H^1(D_+)$, which contradicts the logarithmic singularity of the Green's function at y. Indeed, the derivative of $G_{\mathrm{i}}(\cdot, y)$ has a singularity of order one at y, which is not square integrable. □

It is interesting to observe that in [37] the authors show that the condition $k(f_+ - f_-) < \pi$ implies that a periodic Dirichlet scattering surface is uniquely determined by the scattered field corresponding to one incident plane wave measured on some measurement line above the interface. Hence, if the amplitude of the structure is small enough, then one incident field is sufficient to determine the structure. Our condition $0 < kf_+ < \sqrt{2}$ is somewhat similar to that: for f_+ small enough compared to the wavenumber, the Factorization method works due to coercivity of the middle operator.

Unfortunately, we cannot prove the latter theorem in case that the wavenumber is large compared to the surface elevation of the rough surface Γ. This is especially unsatisfactory since, following the common rule of thumb, inverse scattering techniques should produce better results for higher wavenumbers, see for instance the study in [19] for the linear sampling method. The crucial analytical problem for the inverse rough surface scattering problem is of course the unboundedness of the obstacle, which makes standard compactness arguments impossible. One way to obtain compactness properties in unbounded domains is to use weighted spaces. This is also motivated by the fact that the incident fields in our method are rapidly decaying in x_1 direction. However, there are a couple of arguments preventing a Factorization method in weighted spaces and for the convenience of the reader we sketch some of them in the next section.

Of course, it is again an interesting question how one can approximate the range of $\widetilde{N}$ when the physical data operator N is known. Concerning this topic, we just comment that, due to coercivity of the single layer operator, an approximation result similar to Proposition III.12 from the last chapter can be shown.

For the special case where the rough surface Γ is asymptotically a straight line, we can extend the above Factorization method to all wavenumbers. The reason is that for this special case compact embeddings of unbounded Sobolev spaces hold true: Let $f \in C^1\big([0,\infty)\big)$ be positive and nonincreasing with bounded derivative such that for all $\varepsilon > 0$

$$\lim_{s \to \infty} \frac{f(s+\varepsilon)}{f(s)} = 0.$$

Then [1, Theorem 6.52, Example 6.53] states that for the domain $\Omega = \{(x_1, x_2) : 0 < x_1, 0 < x_2 < f(x)\} \subset \mathbb{R}^2$ the imbedding $H^1(\Omega) \hookrightarrow L^2(\Omega)$ is compact. Motivated by this compact embedding result, assume the rough surface Γ is given by $f \in C^{3,\alpha}(\mathbb{R})$, $\alpha \in (0,1)$, which satisfies the following conditions,

$$\begin{aligned} &\bullet \text{ there is } R > 0\text{: } f'(s) \leq 0 \text{ for } s > R \text{ and } f'(s) \geq 0 \text{ for } s < -R, \\ &\bullet \text{ for all } \varepsilon > 0 \text{ it holds } \lim_{s\to\infty} \frac{f(\pm(s+\varepsilon))}{f(\pm s)} = 0, \\ &\bullet \text{ there are } f_\pm, C:\ 0 < f_- \leq f(s) \leq f_+,\ |f'(s)| \leq C \text{ for all } s \in \mathbb{R}. \end{aligned} \tag{IV.20}$$

The surfaces Γ determined by some f which satisfies the latter assumptions can be considered as a natural extension of the situation considered in the previous chapter, namely, we consider here perturbations of a straight line which are concentrated in some bounded region. The smoothness assumption $f \in C^{3,\alpha}(\mathbb{R})$ is merely necessary to obtain a compactness statement for the difference of two single layer operators later on.

COROLLARY IV.6. *Let the rough surface* $\Gamma = \{(x_1, f(x_1)) : x_1 \in \mathbb{R}\} \subset \mathbb{R}^2$ *be given by a function* $f \in C^{3,\alpha}(\mathbb{R})$ *which satisfies* (IV.20). *Then the embedding* $H^1(D_-) \hookrightarrow L^2(D_-)$ *is compact.*

Proof. For each of the three domains $D_-^1 = \{x \in D_- : x_1 < -R\}$, $D_-^2 = \{x \in D_- : -R-1 < x_1 < R+1\}$ and $D_-^3 = \{x \in D_- : x_1 > R\}$ the embedding of H^1 into L^2 is compact. Hence, let $(u_n)_{n\in\mathbb{N}} \subset H^1(D_-)$ be a bounded sequence. Then there is a subsequence which converges in $L^2(D_-^1)$. From this subsequence we can successively extract further subsequences converging in $L^2(D_-^2)$ and $L^2(D_-^3)$. This final sequence converges in $L^2(D_-)$. □

Now we can state a Factorization method valid for all frequencies in the special case of a rough surface which is asymptotically a flat surface.

THEOREM IV.7. *Let the rough surface* $\Gamma = \{(x_1, f(x_1)) : x_1 \in \mathbb{R}\} \subset \mathbb{R}^2$ *be given by a function* $f \in C^{3,\alpha}(\mathbb{R})$ *which satisfies* (IV.20). *Then the near field operator* $\widetilde{N}$ *has a factorization*

$$\widetilde{N} = -\widetilde{H}^* S^{-1} \widetilde{H},$$

where $\widetilde{H} : L^2(M) \to H^{-1/2}(\Gamma)$ *is compact and injective. Additionally,* $S^{-1} : H^{-1/2}(\Gamma) \to H^{1/2}(\Gamma)$ *is a compact perturbation of a coercive operator and* $\operatorname{Im} S^{-1}$ *is negative semidefinite. Denote by* $(\lambda_j, \phi_j)_{j\in\mathbb{N}}$ *an eigensystem of the positive, compact and selfadjoint operator*

$$\widetilde{N}_\sharp : H^{-1/2}(M) \to H^{1/2}(M), \qquad \widetilde{N}_\sharp = |\operatorname{Re}\widetilde{N}| + \operatorname{Im}\widetilde{N}.$$

Then a point $y \in U_0$ *belongs to* D_- *if and only if*

$$\sum_{j=1}^{\infty} \frac{\left|\langle G_{\mathrm{i}}(\cdot, y), \phi_j\rangle_{L^2(M)}\right|^2}{\lambda_j} < \infty. \tag{IV.21}$$

Proof. We choose k_0 so small that S_{k_0} is coercive between $H^{\pm 1/2}(\Gamma)$ and denote the single layer operator for this wavenumber by S_{k_0}. Splitting the single layer operator in two terms, $S = S_{k_0} + (S - S_{k_0})$, we note that Theorem IV.5 implies that we merely have to show that $S - S_{k_0} : H^{-1/2}(\Gamma) \to H^{1/2}(\Gamma)$ is compact and that S is injective. Indeed, in this case the coercivity of S_{k_0} implies by the Fredholm alternative [63, Theorem 2.27] that $S : H^{-1/2}(\Gamma) \to H^{1/2}(\Gamma)$ is boundedly invertible and the inverse is a compact perturbation of a coercive operator with non-negative imaginary part by the same arguments as in the proof of Theorem IV.5.

Let us first show that S is injective: Assuming that $S\phi = 0$ we obtain from Theorem IV.1 that the potential $\mathrm{SL}\,\phi$ vanishes in D and from [85, Lemma A.2] that $\mathrm{SL}\,\phi$ vanishes in D_-. The jump relations (IV.12) for the single layer potential yield that ϕ vanishes.

We finally show compactness of $S - S_{k_0} : H^{-1/2}(\Gamma) \to H^{1/2}(\Gamma)$ by proving that $\mathrm{SL} - \mathrm{SL}_{k_0}$ maps $H^{-1/2}(\Gamma)$ boundedly into $H^2(D_-)$. Given this result, assume that ϕ_n is a bounded sequence in $H^{-1/2}(\Gamma)$. Then $u_n := \mathrm{SL}\,\phi_n - \mathrm{SL}_{k_0}\,\phi_n$ is bounded in $H^2(D_-)$ and the gradients ∇u_n belong to the (vector-valued) space $H^1(D_-)^2$ and hence, by Corollary IV.6, there is $v \in L^2(D_-)^2$ such that $\nabla u_n \to v$ in $L^2(D_-)$. Extracting a further subsequence we can also assume by Corollary IV.6 that $u_n \to u$ in $L^2(D_-)$. Next we show that $v = \nabla u$ and hence that $u_n \to u$ in $H^1(D_-)$.

For all $n \in \mathbb{N}$ it holds that

$$\int_{D_-} u_n \nabla\phi \,\mathrm{d}x = -\int_{D_-} \nabla u_n \phi \,\mathrm{d}x, \qquad \text{for all } \phi \in C_c^\infty(D_-).$$

The left hand side tends to $\int_{D_-} u \nabla \phi \, \mathrm{d}x$ by the Cauchy-Schwarz inequality, whereas the right hand side tends to $-\int_{D_-} v\phi \, \mathrm{d}x$, proving that

$$\int_{D_-} u \nabla \phi \, \mathrm{d}x = - \int_{D_-} v \phi \, \mathrm{d}x.$$

This implies by the definition of weak differentiability that $v \in L^2(D_-)$ is the weak gradient of u. Consequently, $u_n \to u$ in $H^1(D_-)$ and the trace theorem [63, Theorem 3.37] from $H^1(D_-)$ into $H^{1/2}(\Gamma)$ implies that $u_n|_\Gamma = S\phi_n - S_{k_0}\phi_n \to u|_\Gamma$ in $H^{1/2}(\Gamma)$. Hence, the operator $S - S_{k_0} : H^{-1/2}(\Gamma) \to H^{1/2}(\Gamma)$ is compact.

It remains to show that $\mathrm{SL} - \mathrm{SL}_{k_0} : H^{-1/2}(\Gamma) \to H^2(D_-)$ is bounded. This can be done using the same localization technique as in the proof of the mapping properties of the single layer operator in the end of Section IV-2. Splitting the kernels of the integral operators in a local and a global part as in (IV.7), we note that the difference of the local parts is again locally smoothing. Due to [43, Theorem 2.20] and our smoothness assumption on the rough surface function $f \in C^{3,\alpha}(\mathbb{R})$ we see that the transformed operators $\hat{S}$ and $\hat{S}_{k_0}$ satisfy

$$\left\| (\hat{S} - \hat{S}_{k_0})(\chi_n^2 \phi) \right\|_{H^{s+2}(K_n)} \leq C(\chi^2) \|\chi_n^2 \phi\|_{H^s(\mathbb{R})}, \qquad s \in \{-1, 0\}, \tag{IV.22}$$

where we used the cut-off functions χ_n^2 from Section IV-2 and the bounded set K_n defined in (IV.6). This allows to show the described mapping property of $\mathrm{SL} - \mathrm{SL}_{k_0}$ by the same steps as in Section IV-2. □

We finish this section with some numerical experiments. In the below Figures IV.3–IV.8 we consider a rough surface with profile given by

$$f(x_1) = \begin{cases} 2\sin^2(x_1) \exp\big(-x_1^2/20 - (100 - x_1^2)^{-1} \big) + 3, & |x_1| \leq 10, \\ 2\exp\big(-x_1^2/20 \big) + 3, & |x_1| > 10. \end{cases}$$

This surface is exponentially localized around zero and meets the requirements of assumption (IV.20). Thus, by Theorem IV.7, a Factorization method is theoretically justified independent of the wavenumber. In Figure IV.4, the rough surface is plotted in blue, and also the measurement line is plotted in blue above the surface. The measurement is about half a wavelength long, centered at $x_1 = 0$, and is nearly two wavelengths away

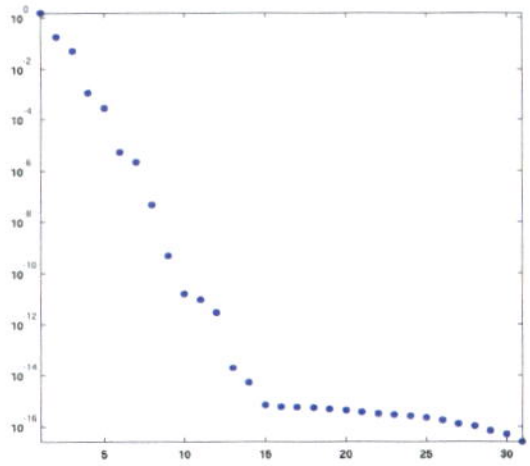

Figure IV.3: Magnitude of eigenvalues of the approximation to $F_\sharp$

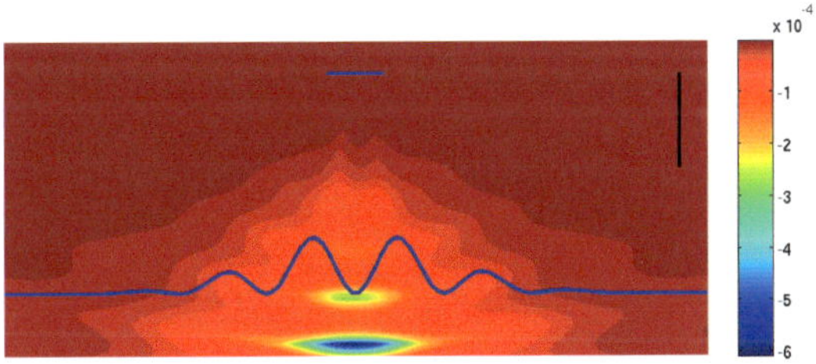

Figure IV.4: $T_{31} = 15$ no artificial noise.

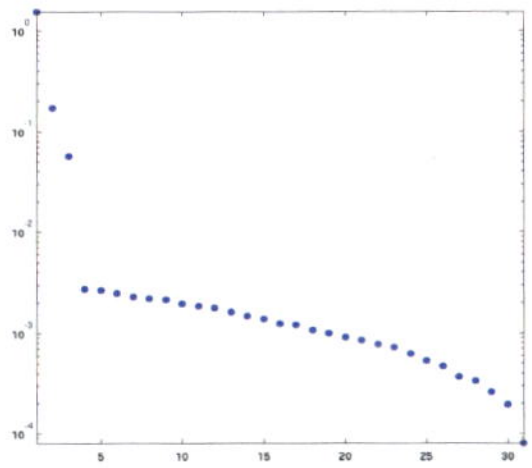

Figure IV.5: Magnitude of eigenvalues of the approximation to $F_\sharp$, one percent artificial noise.

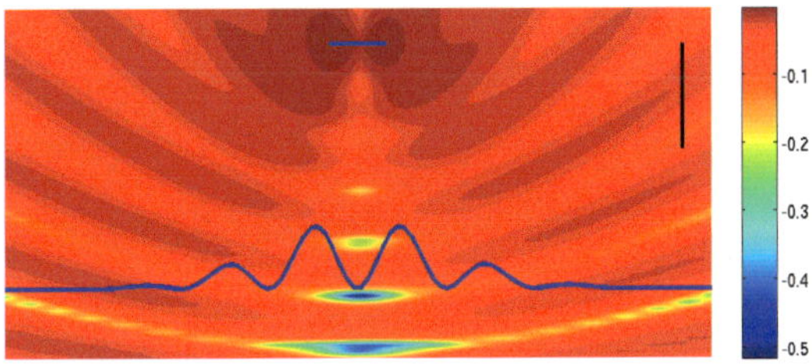

Figure IV.6: $T_{31} = 4$, one percent artificial noise.

from the rough surface. On the measurement line we place 31 equidistant point sources and measurement points to approximate the near field operator N defined in (IV.15). Note that we do not use conjugate point sources for the data operator here and in the following. Using conjugate incident point sources did not improve the overall quality of the reconstructions in our experiments for this rough surface setting. The data for our experiments are computed using an integral equation method as described in [64]. Note that the wavelength used in the direct computations is indicated in the upper right corner of the plots by a vertical black line.

The plot in Figure IV.4 shows the reciprocal of the Picard series (IV.21) truncated at $N = 15$, where no noise has been added to the synthetic data. A glance at Figure IV.3 shows that roughly the first 15 eigenvalues

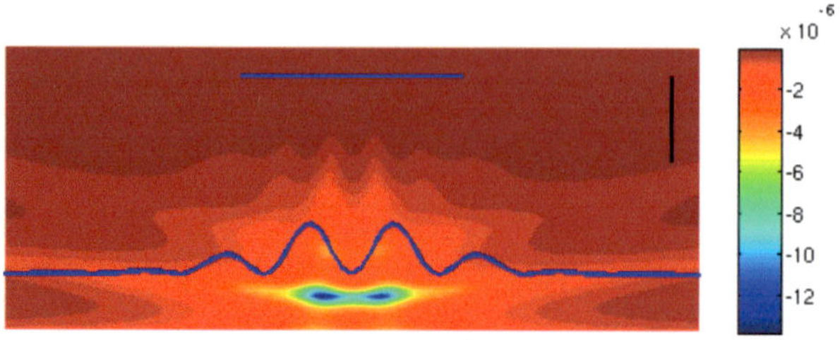

Figure IV.7: $T_{61} = 24$, no artificial noise, large measurement line.

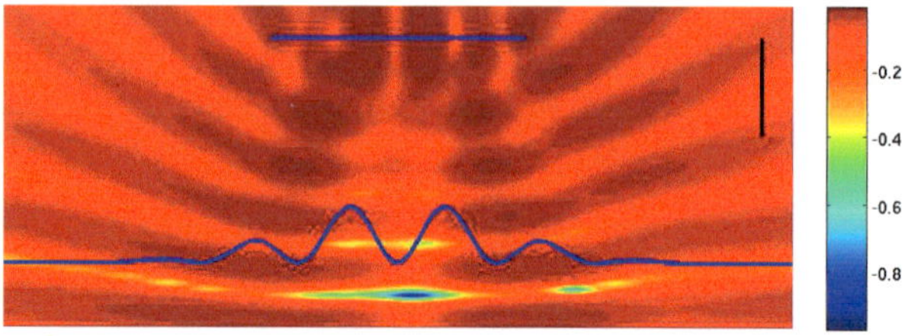

Figure IV.8: $T_{61} = 8$, one percent artificial noise, large measurement line.

of this approximation to $F_\sharp$ seem to be above the noise level, which is of the order of $1 \cdot 10^{-15}$. The plot in Figure IV.4 shows the location of the two big peaks of the surface graph, but is not able to distinguish the two peaks. In Figures IV.5 and IV.6, the same situation is investigated for perturbed data, where one percent relative artificial noise, measured in the Frobenius norm, has been added to the data. Now, all but the largest four eigenvalues are definitely completely perturbed. Plotting the reciprocal of the Picard series shows again the rough location of the peaks of the surface, but now parabola-shaped artefacts appear in a kind of wave-like structure. When we stretch the measurement line and moreover use 61 point sources, see Figures IV.7 and IV.8, the quality of the reconstruction increases. When no artificial noise is added to the synthetic data, the reconstruction even makes the two peaks of the surface visible. Here we truncated the series at $N = 24$ after having a look at the spectrum of the data matrix, which is not shown here. Adding again one percent relative artificial noise to the data, the location of the two peaks is still found, and the artefacts seem to be less dominant than for the case of a small measurement aperture.

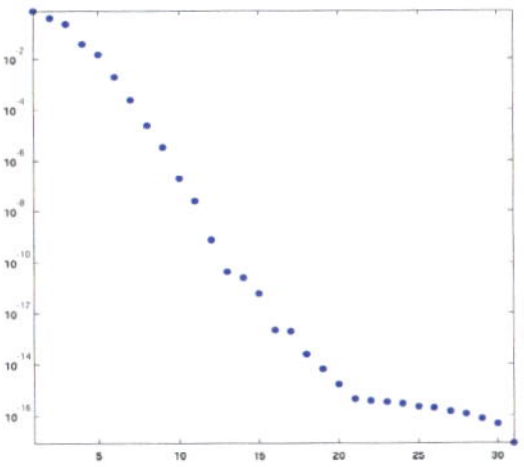

Figure IV.9: Magnitude of eigenvalues of the approximation to $F_\sharp$

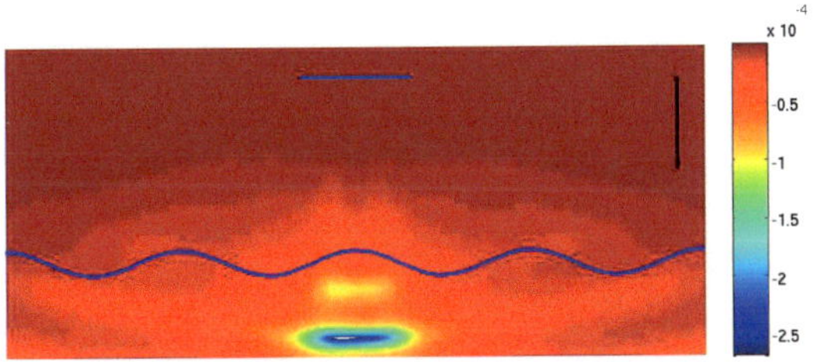

Figure IV.10: $T_{31} = 15$ no artificial noise.

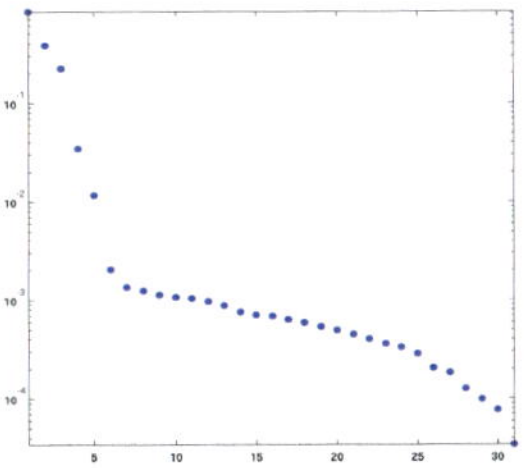

Figure IV.11: Magnitude of eigenvalues of the approximation to $F_\sharp$, one percent artificial noise.

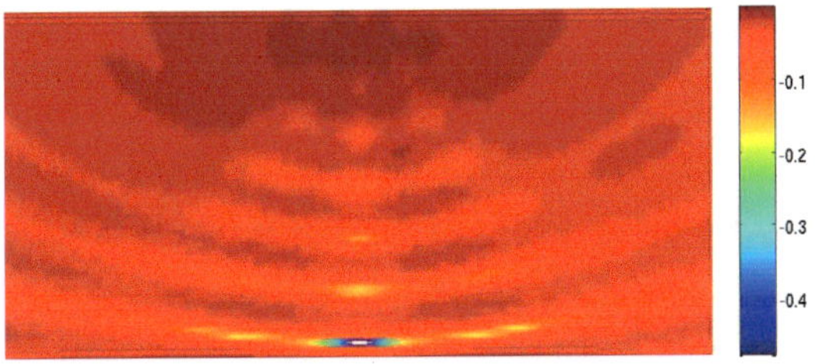

Figure IV.12: $T_{31} = 6$, one percent artificial noise, same scattering surface as in Figure IV.10.

In Figures IV.9–IV.12 we carry out the same experiments as in the last paragraph, but this time for a periodic surface, given by the graph of the function $f(x_1) = \cos(x_1) + 3$. The wavelength and the height of the measurement surface are the same as for the experiments above; again, we place 31 point sources on the measurement surface. Note that the condition $kf_+ < \sqrt{2}$, which we used to set up a Factorization method for this non-local inverse problem, is *not* satisfied. In Figure IV.7 we observe that plotting the reciprocal of the Picard series truncated at $N = 15$ makes the peak of the surface centered around $x_1 = 0$ visible, but none of the neighboring peaks. We add one percent of artificial noise and truncate the

Picard series after the six first terms, which seem to be credible according to the plot of the eigenvalues of the approximation to $F_\sharp$ in Figure IV.11. The reconstruction shows less quality than in the noise-free case, but still the yellow parts of the plot indicates the position of the central peak of the surface. In Figure IV.13 it is demonstrated that a longer measurement line (with 61 point sources on this line) improves the reconstruction also in the noisy case considerably.

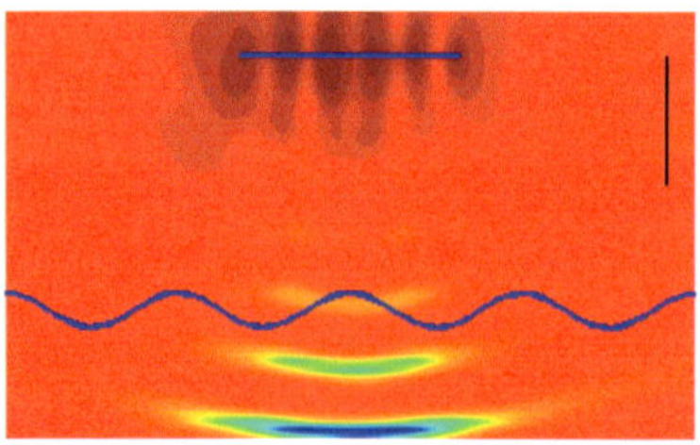

Figure IV.13: $T_{61} = 12$, one percent artificial noise.

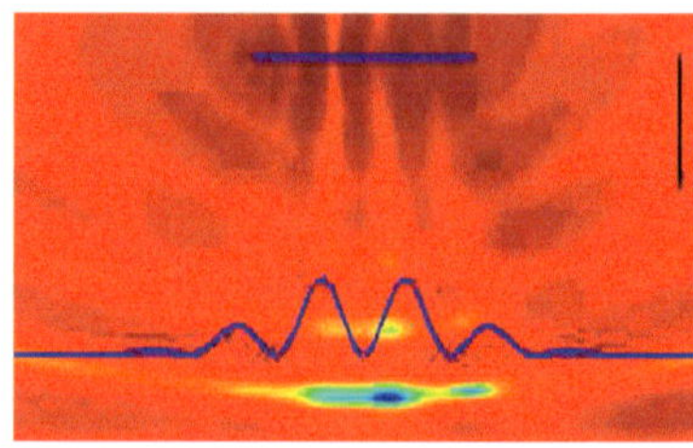

Figure IV.14: noise subspace technique, one percent artificial noise.

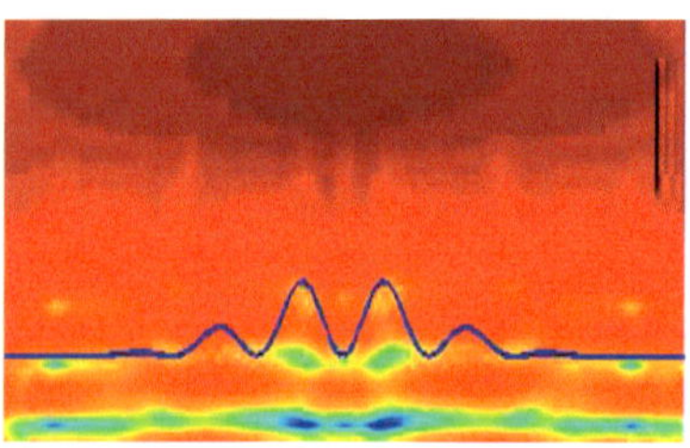

Figure IV.15: Synthetic aperture reconstruction, no artificial noise.

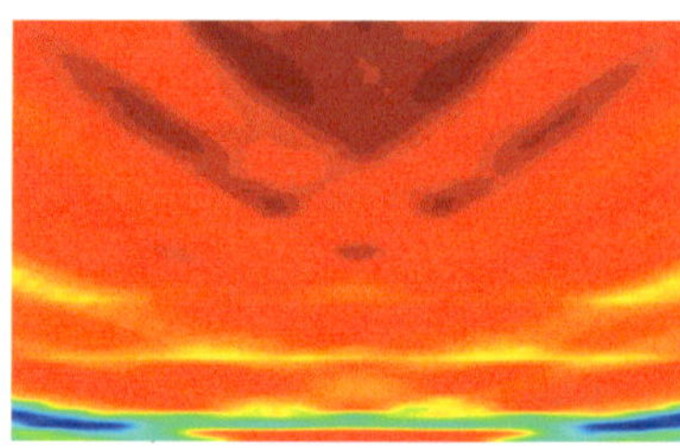

Figure IV.16: Synthetic aperture reconstruction, one percent artificial noise.

To finish the numerical experiments, we show several examples which indicate further directions of research. First, in Figure IV.14 we plot the inverse of the Picard series, which merely consists of five terms corresponding to the five *smallest* eigenvalues. Since the data operator has been perturbed by one percent relative artificial noise, these eigenvalues and also the corresponding eigenspaces have nothing to do with the true eigenvalues and eigenspaces of $F_\sharp$. However, the reconstruction is surprisingly good, even comparable to the reconstruction done in the "standard" way in Figure IV.7 where no artificial noise has been added.

Second, in Figures IV.15 and IV.16 we synthesized the plots by averaging 16 reconstructions obtained from different measurement lines which are all about half a wavelength long. These measurement lines are disjoint and their union covers the entire width of the picture; their vertical position has been the same as on all other plots shown here. On each of the short measurement lines we placed 15 point sources and measurement points and obtained a reconstruction in the same fashion as above. Afterwards, we averaged the 16 reconstructions, which results in a picture shown in Figure IV.15. The technique we use here is connected to synthetic aperture radar imaging [18]. It produces a good reconstruction of the entire rough surface in Figure IV.15, where no artificial noise was added. When we add noise, however, then the parabola-shaped artefacts which already appeared in the examples above make an interpretation of the picture in Figure IV.16 difficult (the rough surface used for this picture is the same as in Figure IV.12). However, appearance of such artefacts is known a-priori and it might be possible to get rid of those in some automatic way by an imaging procedure. If this can be done, this "synthetic aperture" imaging might be an interesting option to obtain global reconstructions of the rough surface.

IV-5. Why Weighted Spaces Do Not Help.

We saw in the last section that Factorization methods for large frequencies fail since the single layer operator S on Γ is no longer coercive. The usual trick in this situation is to split $S = S_{k_0} + (S - S_{k_0})$ where S_{k_0} is the coercive single layer operator for small wavenumber k_0. For bounded obstacles, the difference $S - S_{k_0}$ is compact, however, for the unbounded obstacle Γ this is no longer true, since embeddings of Sobolev spaces on Γ of different order fail to be compact. However, for certain weighted spaces on unbounded domains, one obtains compactness of such embeddings. Therefore one might hope to construct a Factorization method for large frequencies using weighted spaces.

The crucial limitation for a Factorization method in weighted spaces is the following: When we choose X to be the image space of the operator

L in the factorization $\widetilde{N} = -LS^*L^*$, then we necessarily have to consider S^* as an operator from X to X^* where X^* is the dual of X with respect to the pivot space $L^2(\Gamma)$. Indeed, for any other choice of image space for S^*, L and L^* will no longer be adjoint to each other. However, for X being a weighted Sobolev spaces with polynomial weight function, the middle operator between X and X^* is either compact or unbounded.

For the positive and differentiable weight function

$$\rho_\lambda : \mathbb{R}^2 \to \mathbb{R}, \qquad \rho_\lambda(x) = (1 + x_1^2)^{\lambda/2},$$

and a set $\Omega \subset \mathbb{R}^2$ which is a domain or a surface in $\mathbb{R}^2$, we introduce the weighted Sobolev spaces

$$H^s_\lambda(\Omega) := \{u \in \mathcal{D}'(\Omega) : \ \rho_\lambda u \in H^s(\Omega)\}, \qquad \|u\|_{H^s_\lambda(\Omega)} = \|\rho_\lambda u\|_{H^s(\Omega)}.$$

The dual space of $H^s_\lambda(\Gamma)$ for the usual inner product of $L^2(\Gamma)$ is $H^{-s}_{-\lambda}(\Gamma)$. This follows from the usual duality relation of the corresponding unweighted spaces, as

$$|\langle \phi, \psi \rangle| = \left| \langle (1 + |x_1|^2)^{\lambda/2} \phi, (1 + |x_1|^2)^{-\lambda/2} \psi \rangle \right| \leq \|\phi\|_{H^s_\lambda(\Gamma)} \|\psi\|_{H^{-s}_{-\lambda}(\Gamma)},$$

for $\phi \in H^s_\lambda(\Gamma)$ and $\psi \in H^{-1/2}_{-\lambda}(\Gamma)$. It is furthermore possible to show that the outer operator L^* of the Factorization is bounded from $L^2(M)$ into $H^{-s}_\lambda(\Gamma)$ for $\lambda < 1$

Lemma IV.8. *L^* is bounded and injective from $L^2(M)$ to $H^{-1/2}_\lambda(\Gamma)$ for $\lambda < 1$.*

Proof. The adjoint of L is given by

$$L^* \phi = \frac{\partial}{\partial \nu} \mathrm{SL}_M \, \phi \Big|_\Gamma, \quad L^2(M) \to H^{1/2}(\Gamma),$$

where SL_M denotes the single layer potential on M for the Dirichlet Green's function $w(\cdot, y)$ of D, which has been introduced in Chapter III.

Indeed, setting $v = \mathrm{SL}_M\,\phi$, due to $v|_\Gamma = 0$ Green's formula implies

$$\begin{aligned}\langle L\psi, \phi\rangle_{L^2(M)} &= \langle L\psi, [v]_M\rangle_{L^2(M)}\\ &= \int_\Gamma u\frac{\partial \overline{v}}{\partial \nu}\,\mathrm{d}s + \int_{D_h} \left(\nabla u \nabla \overline{v} - k^2 u\overline{v}\right)\,\mathrm{d}x - \int_{\Gamma_h} u\frac{\partial \overline{v}}{\partial \nu}\,\mathrm{d}s\\ &= \int_\Gamma u\frac{\partial \overline{v}}{\partial \nu}\,\mathrm{d}s + \int_{D_h} \left(\nabla u \nabla \overline{v} - k^2 u\overline{v}\right)\,\mathrm{d}x - \int_{\Gamma_h} \overline{v} T u\,\mathrm{d}s\\ &= \int_\Gamma \psi\frac{\partial \overline{v}}{\partial \nu}\,\mathrm{d}s.\end{aligned}$$

Using the Cauchy-Schwarz inequality, we find

$$\begin{aligned}\|L^*\psi\|^2_{L^2_\lambda(\Gamma)} &\le \int_\Gamma (1+|x_1|^2)^\lambda \int_M |w(x-y)|^2\,\mathrm{d}y\,\mathrm{d}x \|\psi\|^2_{L^2(M)}\\ &\le \int_M \int_\Gamma (1+|x_1|^2)^\lambda\, |w(x-y)|^2\,\mathrm{d}x\,\mathrm{d}y \|\psi\|^2_{L^2(M)}\\ &\le C\int_M \int_\Gamma \frac{(1+|x_1|^2)^\lambda}{1+|x-y|^3}\,\mathrm{d}x\,\mathrm{d}y\,\|\psi\|^2_{L^2(M)} \le C\|\psi\|^2_{L^2(M)},\end{aligned}$$

since M is compact and $y \mapsto \int_\Gamma (1+|x_1|^2)^\lambda (1+|x-y|)^{-3}\,\mathrm{d}x$ is uniformly bounded on M. □

Now we can consider the factorization between the spaces $L^2(M)$ and $H^{\pm 1/2}_{\pm\lambda}(\Gamma)$ for $\lambda \in (1/2, 1)$. However, the single layer operator S cannot expected to be bounded from $H^{-1/2}_{-\lambda}(\Gamma)$ into $H^{1/2}_{\lambda}(\Gamma)$: there is no reason why S should map increasing functions boundedly into decaying ones and at least for $|\lambda| < 1/2$ the technique from Section IV-2 shows that S is bounded from $H^{-1/2}_{\lambda}(\Gamma)$ into $H^{1/2}_{\lambda}(\Gamma)$. On the other side, S is compact between $H^{-1/2}_{\lambda}(\Gamma)$ into $H^{1/2}_{-\lambda}(\Gamma)$ for $\lambda > 0$, since it is already bounded from $H^{-1/2}_{0}(\Gamma)$ into $H^{1/2}_{0}(\Gamma)$ and the embedding of $H^{1/2}_{0}(\Gamma)$ into $H^{1/2}_{-\lambda}(\Gamma)$ is easily seen to be compact. This compactness property is independent of the wavenumber. We conclude that for the given family of weighted spaces, there is no hope to improve the wavenumber range of validity of the Factorization method in Theorem IV.5.

List of Figures

List of Symbols

Chapter I

Chapter II

k	wavenumber, page 46
L	data-to-pattern operator, page 60
M_{D}	middle operator, page 64
N	near field operator, page 64
S, K, $\widetilde{K}$, T	boundary integral operators, page 55
S_{D}, K_{D}, $\widetilde{K}_{\mathrm{D}}$, T_{D}	boundary integral operators for Dirichlet Green's function, page 55

Chapter III

Γ	rough surface, page 87
Γ_h	line at height h, page 87
$\mathcal{C}$	contamination on the rough surface, page 93
$\widetilde{H}$	Herglotz operator, page 106
$\widetilde{H}^s(\mathcal{C})$	Sobolev space on the boundary, page 94
$\widetilde{N}$	near field operator, page 105
$C^{1,1}$	Bounded Lipschitz continuous functions with bounded Lipschitz continuous derivative, page 87
D	domain, page 87
D^+	complement of $\overline{D_-}$ in D, page 93
D_-	domain between the contamination and the rough surface, page 93
D_h	truncated domain, page 87
D_h^+	$D_h^+ = D_h \cap D_+$, page 93
G	Dirichlet Green's function for the half space, page 88
H	Herglotz operator, page 109
$H^s(\mathcal{C})$	Sobolev space on the boundary, page 94
N	near field operator, page 108
P_δ	Tikhonov regularization operator, page 110
S	single layer operator, page 99
U_h	half space of height h, page 87
V, $\widetilde{V}$	evaluation operators, page 109
w	Dirichlet Green's function for D, page 99
L	data-to-pattern operator, page 105

Chapter IV

Index

Bibliography

[1] R. A. Adams and J. J. F. Fournier, *Sobolev spaces*, Academic Press, 2. ed., repr. ed., 2005.

[2] T. Arens, *Scattering by biperiodic layered media: The integral equation approach*, 2008. Habilitation Thesis, Universität Karlsruhe, in progress.

[3] T. Arens and N. Grinberg, *A complete factorization method for scattering by periodic structures*, Computing, 75 (2005), pp. 111–132.

[4] T. Arens, K. Haseloh, and S. N. Chandler-Wilde, *Solvability and spectral properties of integral equations on the real line: I. Weighted spaces of continuous functions*, J. Math. Anal. Appl., 272 (2002), pp. 276–302.

[5] ——, *Solvability and spectral properties of integral equations on the real line: II. L^p-spaces and applications*, J. Int. Equ. Appl., 15 (2003), pp. 1–35.

[6] T. Arens and T. Hohage, *On radiation conditions for rough surface scattering problems*, IMA J. Applied Math., 70 (2005), pp. 839–847.

[7] T. Arens and A. Kirsch, *The factorization method in inverse scattering from periodic structures*, Inverse Problems, 19 (2003), pp. 1195–1211.

[8] A.-S. Bonnet-Bendhia and F. Starling, *Guided waves by electromagnetic gratings and non-uniqueness examples for the diffraction*

problem, Mathematical Methods in the Applied Sciences, 17 (1994), pp. 305–338.

[9] C. Burkard and R. Potthast, *Field reconstructions for 3d rough surface scattering*, in Proceedings of the 8th international conference on mathematical and numerical aspects of waves, University of Reading, 2007, pp. 147–149.

[10] F. Cakoni and D. Colton, *On the mathematical basis of the linear sampling method*, Georgian Math. J., 10 (2003), pp. 95–104.

[11] F. Cakoni and D. Colton, *Qualitative Methods in Inverse Scattering Theory. An Introduction.*, Springer, Berlin, 2006.

[12] F. Cakoni, D. Colton, and H. Haddar, *The linear sampling method for anisotropic media*, J. Comp. Appl. Math., 146 (2002), pp. 285–299.

[13] S. Chandler-Wilde, C. Ross, and B. Zhang, *Scattering by infinite one-dimensional rough surfaces*, Proceedings of the Royal Society A, 455 (1999), pp. 3767–3787.

[14] S. N. Chandler-Wilde and P. Monk, *Existence, uniqueness, and variational methods for scattering by unbounded rough surfaces*, SIAM. J. Math. Anal., 37 (2005), pp. 598–618.

[15] S. N. Chandler-Wilde and C. Ross, *Scattering by rough surfaces: the Dirichlet problem for the Helmholtz equation in a non-locally perturbed half-plane*, Math. Meth. Appl. Sci., 19 (1996), pp. 959–976.

[16] S. N. Chandler-Wilde and B. Zhang, *A uniqueness result for scattering by infinite dimensional rough surfaces*, SIAM J. Appl. Math., 58 (1998), pp. 1774–1790.

[17] A. Charalambopoulus, A. Kirsch, K. A. Anagnostopoulus, D. Gintides, and K. Kiriaki, *The factorization method in inverse elastic scattering from penetrable bodies*, Inverse Problems, 23 (2007), pp. 27–51.

[18] M. Cheney, *A mathematical tutorial on synthetic aperture radar*, SIAM Review, 43 (2001), pp. 301–312.

[19] F. Collino, M. Fares, and H. Haddar, *Numerical and analytical study of the linear sampling method in electromagnetic inverse scattering problems*, Inverse Problems, 19 (2003), pp. 1279–1298.

[20] D. Colton, J. Coyle, and P. Monk, *Recent developments in inverse acoustic scattering theory*, SIAM Review, 42 (2000), pp. 396–414.

[21] D. Colton and A. Kirsch, *A simple method for solving inverse scattering problems in the resonance region*, Inverse Problems, 12 (1996), pp. 383–393.

[22] D. L. Colton and R. Kress, *Inverse acoustic and electromagnetic scattering theory*, Springer, 1992.

[23] J. Elschner and G. Schmidt, *Diffraction of periodic structures and optimal design problems of binary gratings. Part I: Direct problems and gradient formulas*, Math. Meth. Appl. Sci., 21 (1998), pp. 1297–1342.

[24] J. Elschner and M. Yamamoto, *Uniqueness results for an invsere periodic transmission problem*, Inverse problems, 20 (2004), pp. 1841–1852.

[25] L. Evans, *Partial Differential Equations*, American Mathematical Society, 1998.

[26] B. Gebauer, M. Hanke, and C. Schneider, *Sampling methods for low-frequency electromagnetic imaging*, Inverse Problems, 24 (2008), p. 015007.

[27] S. Gebauer, *The factorization method for real elliptic problems*, Z. Anal. Anwend, 25 (2006), pp. 81–102.

[28] D. Gilbarg and N. S. Trudinger, *Elliptic Partial Differential Equations of Second Order*, Springer, 2. ed., rev. 3. print. ed., 1998.

[29] I. C. Gohberg and M. G. Kreĭn, *Introduction to the theory of linear nonselfadjoint operators*, vol. 18 of Transl. Math. Monographs, American Mathematical Society, 1969.

[30] R. Griesmaier, *An asymptotic factorization method for inverse electromagnetic scattering in layered media*, SIAM J. Appl. Math., 68 (2008), pp. 1378–1403.

[31] N. I. Grinberg, *Obstacle localization in an homogeneous half-space*, Inverse Problems, 17 (2001), pp. 1113–1125.

[32] H. Haddar, S. Kusiak, and J. Sylvester, *The convex backscattering support*, SIAM J. Appl. Math., 66 (2006), pp. 591–615.

[33] M. Hanke and M. Brühl, *Recent progress in electrical impedance tomography*, Inverse Problems, 19 (2003), pp. S65–S90.

[34] M. Hanke, F. Hettlich, and O. Scherzer, *The Landweber iteration for the inverse scattering problem*, in Proc. of the 1995 design engineering technical conference, K.-W. W. et. al., ed., vol. 3, Part C, New York, 1995, The American Society of Mechanical Engineers, pp. 909–915.

[35] M. Hanke and B. Schappel, *The factorization method for electrical impedance tomography in the half space*, SIAM J. Appl. Math., 68 (2008), pp. 907–924.

[36] F. Hettlich, *Frechet derivatives in inverse obstacle scattering*, Inverse Problems, 11 (1995), pp. 371–382.

[37] F. Hettlich and A. Kirsch, *Schiffer's theorem in inverse scattering theory for periodic structures*, Inverse Problems, 13 (1997), pp. 351–361.

[38] N. Hyvönen, *Complete electrode model of electrical impedance tomography: approximation properties and characterization of inclusions*, SIAM J. Appl. Math., 64 (2004), pp. 902–931.

[39] N. Hyvönen, *Locating transparent regions in optical absorption and scattering tomography*, SIAM J. Appl. Math., 67 (2007), pp. 1101–1123.

[40] N. Hyvönen, H. Hakula, and S. Pursiainen, *Numerical implementation of the factorization method within the complete electrode model of impedance tomography*, Inverse Problems and Imaging, 1 (2007), pp. 299–317.

[41] T. Kato, *Perturbation theory for linear operators*, Springer, repr. of the 1980 ed., 1995.

[42] Y. Katznelson, *An introduction to harmonic analysis*, Cambridge University Press, 3. ed. ed., 2004.

[43] A. Kirsch, *Generalized boundary value- and control problems for the Helmholtz equation*, 1984.

[44] A. Kirsch, *Diffraction by periodic structures*, in Proc. Lapland Conf. on Inverse Problems, L. Pävarinta and E. Somersalo, eds., Springer, 1993, pp. 87–102.

[45] ——, *Uniqueness theorems in inverse scattering theory for periodic structures*, Inverse Problems, 10 (1994), pp. 145–152.

[46] A. Kirsch, *An introduction to the mathematical theory of inverse problems*, Springer, 1996.

[47] A. Kirsch, *The factorization method for a class of inverse elliptic problems*, Math. Nachr., 278 (2004), pp. 258–277.

[48] ——, *An integral equation for Maxwell's equations in a layered medium with an application to the Factorization method*, J. Integral Equations Appl., 19 (2007), pp. 333–358.

[49] A. Kirsch and N. Grinberg, *The Factorization Method for Inverse Problems*, Oxford Lecture Series in Mathematics and its Applications 36, Oxford University Press, 2008.

[50] A. Kirsch and S. Ritter, *A linear sampling method for inverse scattering from an open arc*, Inverse Problems, 16 (2000), pp. 89–105.

[51] R. Kress, *Ein ableitungsfreies Restglied für die trigonometrische Interpolation periodischer analytischer Funktionen*, Numer. Math., 16 (1971), pp. 389–396.

[52] R. Kress, *Linear integral equations*, Springer, 1989.

[53] R. Kress, *Inverse scattering from an open arc*, Math. Methods Appl. Sci., 18 (1995), pp. 183–219.

[54] R. KRESS AND T. TRAN, *Inverse scattering for a locally perturbed half-plane*, Inverse Problems, 16 (2000), pp. 1541–1559.

[55] S. KUSIAK AND J. SYLVESTER, *The scattering support*, Commun. Pure Appl. Math., 56 (2003), pp. 1525–1548.

[56] M. LASSAS, M. CHENEY, AND G. UHLMANN, *Uniqueness for a wave propagation inverse problem in a half-plane*, Inverse Problems, 14 (1998), pp. 679–684.

[57] A. LECHLEITER, *A regularization technique for the factorization method*, Inverse Problems, 22 (2006), pp. 1605–1625.

[58] A. LECHLEITER, N. HYVÖNEN, AND H. HAKULA, *The factorization method applied to the complete electrode model of impedance tomography*, SIAM J. Appl. Math., 68 (2008), pp. 1097–1121.

[59] C. LINES, *Inverse Scattering by Unbounded Rough Surfaces*, PhD thesis, Department of Mathematical Sciences, Brunel University, 2003.

[60] C. D. LINES AND S. N. CHANDLER-WILDE, *A time domain point source method for inverse scattering by rough surfaces*, Computing, 75 (2005), pp. 157–180.

[61] L. LORENZI, A. LUNARDI, G. METAFUNE, AND D. PALLARA, *Analytic semigroups and reaction-diffusion problems*, tech. rep., International School on Evolution Equations, 2005.

[62] D. R. LUKE AND A. J. DEVANEY, *Identifying scattering obstacles by the construction of nonscattering waves*, SIAM J. Appl. Math., 68 (2007), pp. 271–291.

[63] W. MCLEAN, *Strongly Elliptic Systems and Boundary Integral Operators*, Cambridge University Press, Cambridge, UK, 2000.

[64] A. MEIER, T. ARENS, S. N. CHANDLER-WILDE, AND A. KIRSCH, *A Nyström method for a class of integral equations on the real line with applications to scattering by diffraction gratings and rough surfaces*, J. Int. Equ. Appl., 12 (2000), pp. 281–321.

[65] P. MONK, *Finite Element Methods for Maxwell's Equations*, Oxford Science Publications, Oxford, 2003.

[66] P. Monk and J. Sun, *Inverse scattering using finite elements and gap reciprocity*, Inverse Problems and Imaging, 1 (2007), pp. 643–660.

[67] J. Mujica, *Complex analysis in Banach spaces*, North-Holland, 1986.

[68] J. Nečas, *Les Methodes Directes en Theorie des Equations Elliptiques*, Academia, Prague, 1967.

[69] J.-C. Nédélec, *Acoustic and Electromagnetic Equations*, Springer, New York etc, 2001.

[70] G. N.Watson, *A treatise on the theory of Bessel functions*, Univ. Pr., 2. ed., 1966.

[71] M. Orispää, *On point sources and near field measurements in inverse acoustic obstacle scattering*, doctorial thesis, University of Oulu, Oulu, Finland, 2002.

[72] R. Petit, ed., *Electromagnetic theory of gratings*, Springer, 1980.

[73] R. Potthast, J. Sylvester, and S. Kusiak, *A 'range test' for determining scatterers with unknown physical properties*, Inverse Problems, 19 (2003), pp. 533–547.

[74] A. Rieder, *On the regularization of nonlinear ill-posed problems via inexact Newton iterations*, Inverse Problems, 15 (1999), pp. 309–327.

[75] W. Rudin, *Real and complex analysis*, McGraw-Hill, 2. ed., 1974.

[76] W. Rudin, *Functional Analysis*, McGraw-Hill, 2. ed. ed., 1991.

[77] J. Saranen and G. Vainikko, *Periodic integral and pseudodifferential equations with numerical approximation*, Springer, 2002.

[78] H. Schon, *Über die Streuung akustischer Wellen in inhomogenen Medien mit Einschlüssen*, PhD thesis, Universität Karlsruhe (TH), 2005.

[79] F. Schulz, *On the unique continuation property of elliptic divergence form equations in the plane*, Math. Z, 228 (1998), pp. 201–206.

[80] E. Stephan and W. Wendland, *An augmented Galerkin procedure for the boundary integral equation method applied to two-dimensional screen and crack problems*, Appl. Anal., 18 (1984), pp. 183–219.

[81] L. Svensson, *Sums of complemented subspaces in locally convex spaces*, Arkiv för Mathematik, 25 (1987), pp. 147–153.

[82] G. Vainikko, *The discrepancy principle for a class of regularization methods*, U.S.S.R. Comput. Maths. Math. Phys., 21 (1982), pp. 1–19.

[83] G. Vainikko, *Fast solvers of the Lippmann-Schwinger equation*, in Direct and Inverse Problems of Mathematical Physics, D. Newark, ed., Int. Soc. Anal. Appl. comput. 5, Dordrecht, 2000, Kluwer Academic Publishers, p. 423.

[84] D. Werner, *Funktionalanalysis*, Springer, 5 ed., 2005.

[85] B. Zhang and S. N. Chandler-Wilde, *Integral equation methods for scattering by infinite rough surfaces*, Math. Meth. Appl. Sci., 26 (2003), pp. 463–488.

Made in the USA
Las Vegas, NV
03 November 2026

57420367R00105